Organic Chemistry for Students of Health and Life Sciences

Fourth Edition

Paul W. Groundwater
and
Giles A. Taylor

LONGMAN

Addison Wesley Longman
Addison Wesley Longman Limited,
Edinburgh Gate, Harlow,
Essex CM20 2JE, England
and associated companies throughout the world

© Addison Wesley Longman Limited 1997

The right of Paul W. Groundwater and Giles A. Taylor to be identified as the authors of this work has been asserted by them in accordance with the Copyright, Designs and Patents Act 1988

All rights reserved; no part of this publication may be reproduced, stored in any retrieval system, or transmitted in any form or by any means, electronic, mechanical, photocopying, recording, or otherwise without either the prior written permission of the Publishers or a licence permitting restricted copying in the United Kingdom issued by the Copyright Licensing Agency Ltd, 90 Tottenham Court Road, London W1P 9HE

First published 1971
Second edition 1978
Third edition 1987
Fourth edition 1997
Reprinted 1998

British Library Cataloguing in Publication Data
A catalogue entry for this title is available from the British Library

ISBN 0-582-29765-6

Library of Congress Cataloging-in-Publication Data
A catalog entry for this title is available from the Library of Congress

Set by 40 in 10/12 Times Roman
Produced through Longman Malaysia, VVP

Organic Chemistry for Students of Health and Life Sciences

THIS BOOK BELONGS TO:
Kostoris Library
The Christie Hospital NHS Trust
Manchester
M20 4BX
Phone: 0161 446 3452
Email: library@picr.man.ac.uk

From PWG

To Phyllis, Ellen, Catriona and Amy

Contents

Preface		ix
1	Basic principles: atomic and molecular structure, nomenclature, stereochemistry and mechanism	1
	1.1 Atomic and molecular structure	1
	1.2 Isomerism and nomenclature	13
	1.3 Stereochemistry	16
	1.4 Mechanism	37
	1.5 Summary	52
	Problems	54
2	Simple organic oxygen and sulphur compounds: alcohols, phenols and ethers, and their sulphur analogues	58
	2.1 Alcohols	58
	2.2 Phenols	67
	2.3 Ethers	70
	2.4 Simple sulphur compounds	72
	2.5 Summary	75
	Problems	77
3	Carbonyl compounds: aldehydes and ketones	79
	3.1 Aldehydes and ketones	80
	3.2 Quinones	95
	3.3 Summary	96
	Problems	99
4	Carbonyl compounds: carboxylic acids and their derivatives	102
	4.1 Carboxylic acids	102
	4.2 Carboxylic acid derivatives: esters	112
	4.3 Carboxylic acid derivatives: acyl halides	115
	4.4 Carboxylic acid derivatives: acid anhydrides	116
	4.5 Carboxylic acid derivatives: thioesters	116
	4.6 Carboxylic acid derivatives: amides	117
	4.7 Summary	118
	Problems	120

5	Simple organic nitrogen compounds: amines	122
	5.1 Amines	122
	5.2 Aromatic diazonium salts	131
	5.3 Summary	133
	Problems	135
6	Carbohydrates	137
	6.1 D-Glucose	137
	6.2 D- and L-Sugars	141
	6.3 D-(−)-Fructose	142
	6.4 Reactions of sugars	143
	6.5 Oligosaccharides and polysaccharides	148
	6.6 Enzymic degradation of starch and cellulose	153
	6.7 Summary	155
	Problems	156
7	Amino acids, peptides and proteins	159
	7.1 Amino acids	159
	7.2 Peptides and proteins	164
	7.3 Summary	170
	Problems	171
8	Aromatic compounds, nucleic acids and nucleotide coenzymes	172
	8.1 Aromatic compounds	172
	8.2 Nucleic acids	185
	8.3 Nucleotide coenzymes	189
	8.4 Summary	193
	Problems	195
9	Lipids	196
	9.1 Fatty acids	197
	9.2 Plant and animal waxes	198
	9.3 Depot fats	198
	9.4 Phospholipids	200
	9.5 Lipids and the structure of biological membranes	202
	9.6 Summary	205
	Index	207

Preface

It is approximately 10 years since the production of the third edition of this book and during this time there has been a dramatic change in both the structure and teaching of undergraduate degrees. The changes we have made to the book are intended to reflect the way in which chemistry is now taught and in particular, the adoption in most universities of a modular degree system. In addition, we have re-organised the material into what we believe to be a more sysytematic format, with all the basic concepts (structure, isomerism, and mechanism) covered in the first chapter and the chemistry of the key functional groups in biological systems covered in chapters 2–5. Finally, the chemistry of some of the most important biological molecules is described, in terms of the functional groups they contain, in chapters on; carbohydrates; amino acids, peptides and proteins; aromatic compounds, nucleic acids and nucleotide coenzymes; and lipids. We have also tried to give the text a more up to date look.

The authors would like to thank Miss Edanie Beccano and Mrs Noreen Gray for the typing of the manuscript and all at AWL for their help and patience.

1 Basic principles: atomic and molecular structure, nomenclature, stereochemistry and mechanism

Topics

1.1 Atomic and molecular structure
1.2 Isomerism and nomenclature
1.3 Stereochemistry
1.4 Mechanism
1.5 Summary

1.1 Atomic and molecular structure

The **atom** (the smallest particle of an element retaining the chemical properties of the element) is the basic structural unit of chemical compounds. All matter is composed of atoms, approximately one hundred types being known, most of which occur naturally.

Atoms can combine to form molecules, and the methods and patterns of combination, and the circumstances in which combination occurs, are the concern of the chemist. A knowledge of the structure of atoms is necessary to understand the reasons why reactions occur, and the phenomenon of valency (the combining power of atoms).

1.1.1 Atomic structure

The structure of the atom can be divided into two distinct features. The **nucleus** is a small, dense, positively charged body at the centre of the atom, and almost all the mass of the atom is concentrated in the nucleus. A diffuse zone containing **electrons** (light particles, with unit negative charge) surrounds the nucleus, and it is these electrons that are responsible for the formation of chemical bonds.

The *nucleus* is itself composed of two types of particle:

- **Protons**, heavy particles with unit positive charge;
- **Neutrons**, heavy particles with no electrical charge.

Both the proton and the neutron have masses approximately two thousand times that of the electron, whose contribution to the total mass of the atom is thus extremely small.

The number of protons in the nucleus of an atom determines the total positive charge in the nucleus, and therefore the number of electrons required to produce a neutral atom. The number of protons, which is the same as the atomic number of the element concerned, thus determines the chemistry of the atom, and is the principal source of difference between the atoms of the elements in the periodic table. The number of neutrons in the nucleus is not so significant, since these have very little effect on the chemistry of the atom, and neutrons may be regarded as an optional extra in the construction of the nucleus.

Isotopes are atoms that differ only in the number of neutrons and are thus chemically indistinguishable. Many of the naturally occurring elements are composed of more than one isotope (see Table 1.1) though some, like fluorine and sodium, occur naturally only as one isotope. When the isotopes are described by symbols, as in Table 1.1, the superscript prefix refers to the mass of the nucleus (i.e. the number of protons and neutrons).

The practical value of the relative atomic mass of an element is a mean value determined by the natural proportions of the isotopes (e.g. chlorine is approximately 75 per cent ^{35}Cl and 25 per cent ^{37}Cl and has a practical relative atomic mass of 35.5; similarly bromine, composed of nearly equal proportions of ^{79}Br and ^{81}Br, has a relative atomic mass of 80).

The *electrons* are found outside the nucleus, and occupy a very much larger space. They are often visualised as minute spherical bodies moving in concentric circles, but this concept is quite wrong and very misleading. It is now known that it is quite impossible to treat electrons in a way that implies that their positions at any moment can be defined accurately. All that can be achieved is to define the probability of finding an electron in a region of space.

The electrons of a polyelectronic atom are not all equivalent. They may be divided firstly into groups known as **shells**, which differ greatly in energy, and which can accommodate differing maximum numbers of electrons. The shells are normally described by a quantum number 1, 2, 3, etc., where shell no. 1 has the lowest energy, no. 2 has the next to the lowest energy, and so on. A polyelectronic atom can be built up by taking the nucleus and bringing electrons from an infinite distance (where their energy is arbitrarily defined as zero) to fill the vacant shells in much the same way that a set of drawers might be filled, starting from the bottom. The first two electrons go into the lowest energy shell (no. 1), which is then full. Subsequently the second shell (8 electrons) and third shell (18 electrons) will be filled, and so on until a sufficient

Table 1.1
The composition of some natural isotopes

Name	Symbol	Atomic number	Nucleus No. of protons	No. of neutrons	No. of electrons	Approx. relative atomic mass	% natural abundance
Hydrogen*							
(Protium)	^{1}H	1	1	0	1	1	99.98
(Deuterium)	^{2}H (D)	1	1	1	1	2	0.02
(Tritium)	†H (T)	1	1	1	1	3	
Helium	^{3}He	2	2	1	2	3	1.3×10^{-4}
	^{4}He	2	2	2	2	4	99.99
Lithium	^{6}Li	3	3	3	3	6	7.3
	^{7}Li	3	3	4	3	7	92.7
Sodium	^{23}Na11	11	12	11	23	100	
Potassium	^{39}K	19	19	20	19	39	93.3
	$^{\dagger 40}$K	19	19	21	19	40	0.01
	^{41}K	19	19	22	19	41	6.7
Fluorine	^{19}F	9	9	10	9	19	100
Chlorine	^{35}Cl	17	17	18	17	35	75.4
	^{37}Cl	17	17	20	17	37	24.6
Bromine	^{79}Br	35	35	44	35	79	50.5
	^{81}Br	35	35	46	35	81	49.5
Carbon	^{12}C	6	6	6	6	12	98.9
	^{13}C	6	6	7	6	13	1.1
	^{14}C	6	6	8	6	14	2×10^{-10}
Nitrogen	^{14}N	7	7	7	7	14	99.6
	^{15}N	7	7	8	7	15	0.4
Oxygen	^{16}O	8	8	8	8	16	99.76
	^{17}O	8	8	9	8	17	0.04
	^{18}O	8	8	10	8	18	0.2

*For historical reasons the isotopes of hydrogen have acquired individual names.
† Radioactive isotope.

number of electrons have been added to make an electrically neutral atom (Figure 1.1).

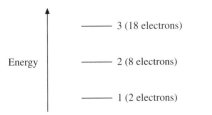

Figure 1.1

Within the electron shells the electrons are not equivalent, as each shell is divided into **orbitals**. Each orbital can accommodate only two electrons, and is distinguished from other orbitals in the same shell by its geometry. An orbital can be regarded as a mathematically defined region of space in which there is a high probability of finding an electron.

We need be concerned here only with the orbitals associated with the first two shells (i.e. found in the first ten elements of the periodic table). These are of two types known as s and p orbitals. s Orbitals are spherically symmetrical

4 Basic principles: structure, nomenclature, stereochemistry and mechanism

about the nucleus, but p orbitals are symmetrical about an axis and a perpendicular plane, both of which pass through the nucleus. The orbitals are generally described by the quantum number of the shell, and the letter designating the type of orbital, e.g. 2s, 2p.

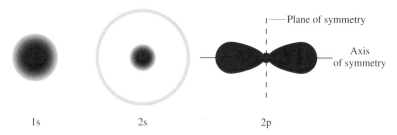

Figure 1.2

The electrons in hydrogen and helium are found in the 1s orbital, this being the only orbital in the first shell and the one responsible for the formation of chemical bonds to hydrogen atoms (Figures 1.2 and 1.3, Table 1.2). As can be

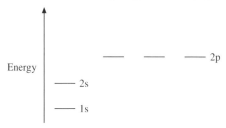

Figure 1.3

seen, the 2s orbital is a similar shape to the 1s but, being more diffuse, it is higher in energy. There are four orbitals available in the second shell, one s type, and three p type differing in the orientation of their axes of symmetry, which are mutually perpendicular (see Figure 1.4). These 2p orbitals are distinguished by the suffixes x, y and z.

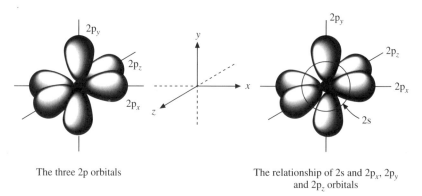

Figure 1.4

More complex atoms (e.g. lithium, neon) can be built up by adding the correct number of electrons to the orbitals in order of increasing energy (*aufbau* principle). In doing this we must take into account two important factors:

Table 1.2
The electronic structure of atoms of the first short period

Atom	Symbol	\multicolumn{5}{c}{No. of electrons in the orbitals}

Atom	Symbol	1s	2s	$2p_x$	$2p_y$	$2p_z$
Hydrogen	H	1				
Helium	He	2				
Lithium	Li	2	1			
Beryllium	Be	2	2			
Boron	B	2	2	1		
Carbon	C	2	2	1	1	
Nitrogen	N	2	2	1	1	1
Oxygen	O	2	2	2	1	1
Fluorine	F	2	2	2	2	1
Neon	Ne	2	2	2	2	2

- **Pauli exclusion principle**. This states that only two electrons can occupy a given orbital, and only then if they have different electron spin, i.e. ↑↓ paired electrons (not ↑↑ or ↓↓).
- **Hund's rule**. This states that if two electrons occupy two or more degenerate (equal in energy) orbitals, such as the 2p orbitals, their mutual repulsion energy will be at a minimum if they have unpaired spins and so occupy different orbitals (see Table 1.2; carbon, nitrogen).

1.1.2 Molecular orbitals and structure

The formation of compounds from elements is achieved by the combination of atoms into groups (**molecules**) whose structure is characteristic of the compound concerned. Some aspects of the formation of molecules and their structures will now be examined.

Valency is the combining power of an element, measured by the number of hydrogen atoms (or their equivalent) with which an atom of the element will combine to form a stable molecule. It is well known that the valency of an element is related to its position in the periodic table in a way that suggests that atoms with unfilled outer shells of electrons attempt to achieve a 'rare gas structure', i.e. a completed outer shell. There are two principal ways of achieving this stable state:

- **Electrovalency** caused by the loss or gain of electrons, to form charged species (**ions**) with complete outer shells.
- **Covalency** in which atoms achieve the equivalent of a rare gas structure by sharing electrons.

Electrovalency is mostly found in compounds involving the elements of groups I, II, VI and VII, since here the gain or loss of no more than two electrons from a cation, or addition to an anion, becomes progressively more difficult, so that true species of the type Al^{3+} or N^{3-} are very rare. Covalency, which avoids the formation of these highly charged species, is thus principally found in the compounds of the central groups of the periodic table (3–15). The production of C^{4+} or C^{4-} is impossible under normal conditions, and so we find that organic chemistry – the chemistry of carbon – is almost entirely the chemistry of covalently bound molecules.

6 Basic principles: structure, nomenclature, stereochemistry and mechanism

Covalency is often taken to refer only to a sharing of two electrons in which one electron comes from each atom. The situation in which a bond is formed by the sharing of two electrons, both of which come from the same atom, is sometimes described as **coordinate valency** and is represented by the symbol A → B where A is supposed to donate both electrons. However, the bond formed in this way is identical to the normal covalent bond except that a separation of charges occurs. This can be seen to arise if the bond is formed in two steps:

- donation of one electron from A: to B giving A$^+$ and B$^-$;
- sharing of one electron each from A$^+$ and B$^-$ to form a covalent bond.

The **semipolar bond** so formed is much better represented as A$^+$—B$^-$ which emphasises both the true covalent character of the link and also the charge relationship between the two atoms.

The formation of bonds

We are now in a position to discuss the formation of covalent bonds, and to understand some of the characteristic features of molecules constructed in this way. The formation of covalent bonds by the sharing of electrons results from the overlapping and interaction of partly filled atomic orbitals. The **molecular orbitals** (bonds) so formed are represented adequately by a simple sum of the geometrical properties of the individual atomic orbitals.

Two s orbitals on different atoms can overlap *in phase* or *out-of-phase*, to give a bonding (σ) or antibonding (σ*) orbital, each of which can accommodate two electrons (Figure 1.5).

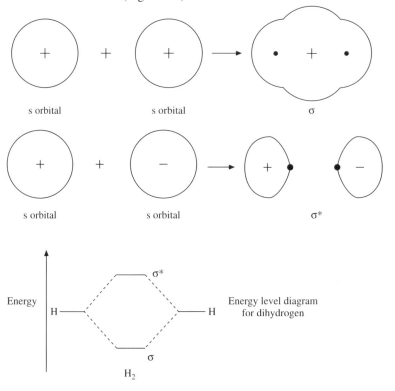

Figure 1.5

Of the *three p orbitals*, one will be different from the others in that it lies along the axis between the two atoms forming a bond. This orbital, e.g. p_z, can thus form a σ molecular orbital with either an s or another p_z orbital.

σ Orbitals (σ bonds) are formed by overlap along the axis between the atoms and, since this is a good overlap, give rise to strong bonds. π Orbitals are formed by sideways overlap above and below the axis between the atoms and, since this is a less efficient overlap, give rise to weaker bonds (Figure 1.6). (Both σ and pσ above are bonding. The corresponding out-of-phase interactions will give the antibonding orbitals, σ* and pσ*.)

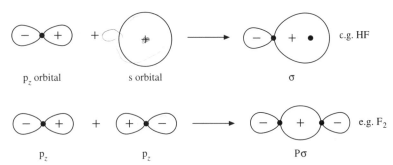

Figure 1.6

For the overlap of a p_x (or p_y) orbital with the s orbital, the in-phase and out-of-phase interactions cancel, giving a non-bonding orbital (Figure 1.7). In

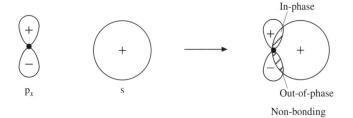

Figure 1.7

addition, the parallel p orbitals on different atoms can overlap sideways (p_x with p_x, p_y with p_y) to form pπ orbitals (Figure 1.8). It should be noted that, for all the bonds shown, there is a finite probability that the electrons will be found outside the internuclear zone.

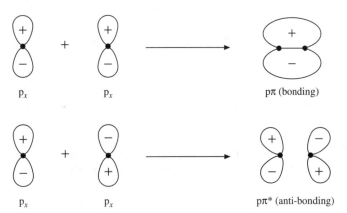

Figure 1.8

8 Basic principles: structure, nomenclature, stereochemistry and mechanism

The three-dimensional structure of molecules

One of the characteristic features of polyatomic molecules is that they have a definite three-dimensional structure, in which the relative positions in space of the covalently linked atoms are fixed. Since bond formation is produced by the interaction of orbitals, which can be highly directional, the most effective bonds will be formed when the relative spatial positions of the atoms are such as to produce the best possible overlap of orbitals. Any distortion of the molecule which, by moving atoms from these most favourable sites, reduces the effectiveness of orbital overlap, will result in a weakening of the bond. Merely by considering the relative orientation of the 2p orbitals, we can make a rough prediction of the geometry of some simple molecules.

In molecules such as NH_3 (**1**) or H_2O (**2**) (Figure 1.9) the N—H or O—H bonds are formed from the 1s orbital of hydrogen and the 2p orbitals of nitrogen or oxygen. We have seen earlier that the three 2p orbitals are mutually perpendicular (p. 4). We might therefore expect that the three N—H bonds or two O—H bonds would also be at right angles, and this is approximately correct, $\angle H—N—H = 107°$, $\angle H—O—H = 104°$. The deviation from 90° is probably caused by mutual repulsion of the positively charged hydrogen nuclei. In H_2S where the H—S bonds are formed from 3p orbitals (similar to 2p in geometry), the hydrogen atoms are further apart owing to the greater size of the sulphur atom, the repulsion between adjacent hydrogen nuclei is therefore less, and in consequence $\angle H—S—H = 93°$.

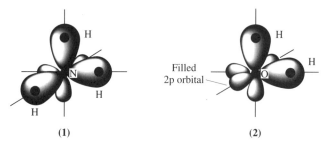

Figure 1.9

Hybridisation of orbitals

We have seen that in compounds of trivalent nitrogen, divalent oxygen and monovalent fluorine p orbitals can be used for bond formation.

The majority of organic compounds are tetravalent and their structure has been shown to be tetrahedral, with all four bonds being identical, e.g. CH_4 or CCl_4. Tetrahedral carbon compounds cannot be adequately explained simply by use of the 2s orbital to form a fourth bond (since this would result in three directed bonds – mutually perpendicular – and one non-directional bond). Pauling and Slater resolved this discrepancy by introducing the concept of hybridisation (mixing) of orbitals. The most effective orbitals for bond formation are those with the best geometry for overlap, and p orbitals have the desired directional shape, whereas s orbitals with spherical symmetry have the least suitable shape. A compromise solution is found to occur with tetravalent atoms. Interaction of the s and p orbitals produces sets of **hybrid orbitals** whose directional properties are somewhat less desirable than pure p orbitals, but very much better than s orbitals. This hybridisation can occur in three ways and the type adopted in any particular case is that which leads to the lowest overall energy (i.e. the strongest bonding).

1. All four orbitals can interact to give four identical hybrid orbitals described as sp³ orbitals (since they are produced by the mixing of one s and three p orbitals). The four sp³ orbitals have the geometry shown in Figure 1.10 (**3**), and are arranged about the nucleus with tetrahedral symmetry (109°) (sp³: 25% s character; 75% p character).

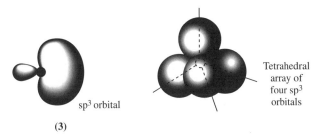

Figure 1.10

(**3**)

When an atom of the first short period forms a compound in which it is attached to four neighbouring atoms by single bonds, sp³ hybrid orbitals are used to form these bonds. The tetrahedral symmetry of the sp³ orbitals about the nucleus means a tetrahedral array of bonds, and all the species of type AX_4 (e.g. BF_4^-, CH_4, CCl_4, $\overset{+}{N}H_4$) are known to have structures in which the X atoms lie at the corners of a regular tetrahedron, with atom A at the centre, and all the angles ∠X—A—X = 109.5° approximately as in (**4**) in Figure 1.11. The overlap potential of the sp³ orbitals is greater than that of either the s or p orbitals, so they are capable of forming stronger bonds than either of the individual atomic orbitals.

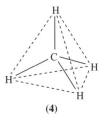 

(**4**)

The structure of CH_4 (methane) Dotted lines show the regular tetrahedron defined by the H atoms

Geometry of the molecular orbital of the C—H bond in methane

Figure 1.11

2. The s orbital and two of the p orbitals can interact to form three sp² orbitals. These are arranged symmetrically in a plane, perpendicular to the unused p orbital (**5**) (Figure 1.12), with an angle of 120° between them (sp²: 33% s character; 66% p character). Once again the overlap potential of the sp² orbitals is greater than that of either of the individual atomic orbitals.

An atom only forms as many hybrid orbitals as it has atoms attached to form strong σ bonds to. Thus, sp² orbitals are used only in the formation of molecules containing double bonds. In a molecule such as ethene, $CH_2\!=\!CH_2$, the two bonds joining the carbon atoms are not identical, and are formed in different ways. The atomic orbitals of carbon are hybridised

Figure 1.12

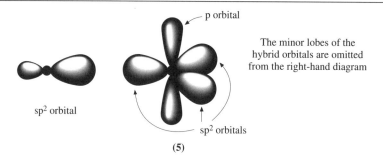

(5)

to form three sp² orbitals and one p orbital, and a framework of single bonds is built up by σ bonds (p. 6) resulting from the overlap of sp² orbitals of carbon and s orbitals of hydrogen. (The molecular orbitals of these σ bonds are illustrated separately, but for convenience are represented by conventional straight lines in Figure 1.13 (**6**).) The remaining two unhybridised p orbitals on adjacent carbon atoms can overlap to give a molecular orbital (bond) consisting of two diffuse electron clouds lying on opposite sides of the C—C axis (**7**). Since this is a sideways overlap, it is a π bond. All double bonds formed by the elements of the first short period (e.g. C=C, C=O, N=N) are constructed in this way and consist of one strong (σ) bond and one weak (π) bond. As would be predicted from such an arrangement, C=C is stronger than C—C (bond dissociation energy 598 kJ mol⁻¹ versus 347 kJ mol⁻¹), but is less than twice as strong.

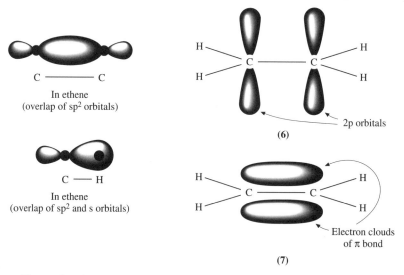

Figure 1.13

The molecular geometry of compounds with double bonds is determined by two factors. In the first place the planar array of sp² orbitals about the nucleus (p. 9) means that the central atom and its three substituent atoms will be coplanar, with bond angles of 120°. In addition, for the p orbitals on adjacent atoms to overlap effectively they must be aligned parallel in space. This means that all the sp² orbitals of the two doubly bonded atoms and all their substituent atoms must be coplanar. Physical measurements confirm this in many cases, e.g. in ethene, all six atoms lie in the same plane.

3. The s orbital and one of the p orbitals can interact to give two sp orbitals. These are arranged linearly (180°), and are perpendicular to the remaining two p orbitals (**8**) (Figure 1.14) (sp: 50% s character, 50% p character). Once again, the sp orbital forms stronger bonds than either the s or p orbitals.

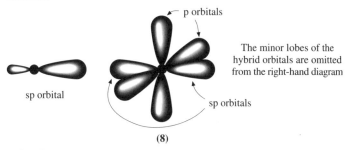

Figure 1.14

In alkynes (or other triply bonded functional groups) the carbon only has two groups to which it can form strong σ bonds to and thus only forms sp orbitals. These are used principally where atoms are joined by triple bonds. In ethyne, H—C≡C—H, the central σ bond skeleton is built of sp orbitals of carbon and s orbitals of hydrogen, which are similar to the corresponding bonds in ethene. The remaining pairs of p orbitals on the carbon atoms (**9**) (Figure 1.15) can overlap to form two π bonds, which together form a cylindrical electron cloud surrounding the C—C axis (**10**). Where atoms of the first short period are joined by triple bonds (e.g. N≡N) the molecular orbitals are constructed in this way. The linear array of sp orbitals about the nucleus, means that the two triply bonded atoms and the substituents are co-linear, as is found to be the case in ethyne and HCN. Thus, a triple bond consists of a strong (σ) bond and two weak (π) bonds and is less than three times as strong as a single bond (812 kJ mol^{-1} versus 347 kJ mol^{-1}).

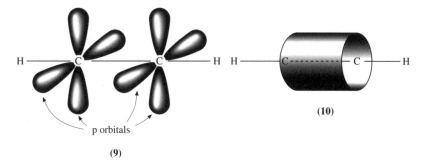

Figure 1.15

Conjugation
It has already been mentioned that the electrons of a σ bond are confined predominantly to a zone between the nuclei of the atoms joined. The electrons of a π bond are, by comparison, much less tightly held, and the electron clouds of the π bond are consequently much more diffuse and can sometimes interact with the electron clouds of adjacent π bonds. This interaction is particularly important in molecules that contain chains of atoms linked alternately by single and double bonds.

12 Basic principles: structure, nomenclature, stereochemistry and mechanism

We will consider the molecule butadiene, $CH_2=CH—CH=CH_2$, whose carbon skeleton is joined by just such an alternate sequence of bonds. Neglecting the CH bonds, which are irrelevant for the present considerations, the basic skeleton (**11**) (Figure 1.16) of the molecule can be represented by

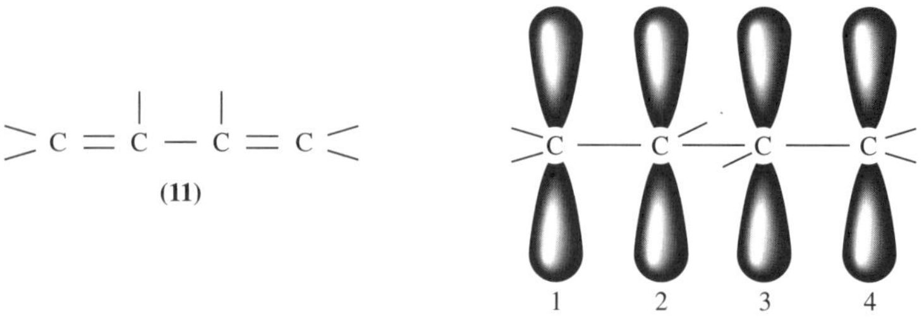

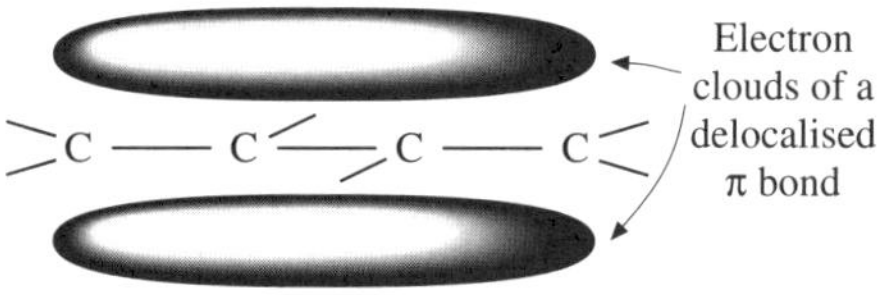

Figure 1.16

(**12**) in which each carbon atom is joined to its neighbours by σ bonds produced by the overlap of atomic sp² orbitals, and each carbon atom has one spare p orbital containing one electron. To form the two double bonds in the conventional representation, the p orbitals on atoms 1 and 2 must overlap, and likewise the p orbitals on atoms 3 and 4. However, the p orbital on atom 2 is equally close to those on atoms 1 and 3 and interaction could occur in either direction. The result of this proximity is that all four p orbitals combine to produce two molecular orbitals which stretch the whole length of the chain of four carbon atoms. Each of these molecular orbitals (π bonds) contains two electrons, and one is illustrated in (**13**).

An alternate sequence of double and single bonds, such as we have just considered, is known as conjugated and will always produce extended ('delocalised') molecular orbitals in which electrons are free to move the whole length of the unsaturated system. In β-carotene, one of the colouring materials of carrots, these molecular orbitals stretch over 22 carbon nuclei and are responsible for the colour of the compound.

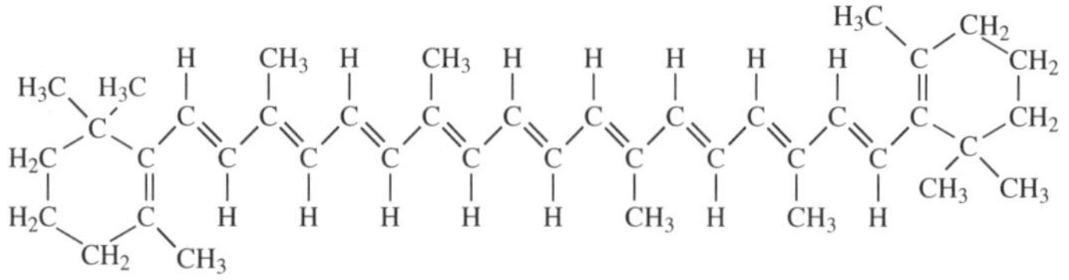

Where double bonds are separated by two or more single bonds, e.g. $CH_2=CH—CH_2—CH=CH_2$, the π bonds are too far apart to interact with each other and no extended molecular orbital can be produced.

Resonance

It is not always possible to represent the structure of a compound adequately by a single, conventional, structural formula compatible with all the properties of the compound, and in these circumstances the correct description often appears to be an average of several conventional structures. A simple example of this is the nitrate ion, which can be represented in three ways:

(14)

To obey the normal rules of valency one of the three oxygen atoms must be joined to the nitrogen atom by double bonds, one by a semipolar (coordinate) bond, and one – bearing a negative charge – by a single covalent bond. There are three different ways of fulfilling these requirements, distinguished by which of the three oxygen atoms is doubly bonded (the other two oxygen atoms turn out to be indistinguishable on account of the structure of the semipolar bond), but none of these is correct as it is known from X-ray examination of nitrates that all three N—O bonds are of identical length, and therefore electronically identical. In view of this we can say only that the true structure of the NO_3^- ion is an average of these three structures, which cannot be represented adequately by conventional symbols. The nitrate ion is described as a **resonance hybrid** (= average) of the three **canonical structures** (orthodox structures). Where it is required to indicate that resonance occurs between canonical structures (*which can differ only in the distribution of electrons*), the double-headed arrow is employed as shown in (**14**). Species, to which no single adequate structure can be assigned, are described as **mesomeric** (e.g. the nitrate ion is a mesomeric anion).

The frequent use of a pendulum analogy to illustrate the concept of resonance creates the totally false idea that resonance is a rapid interchange between a number of extreme structures. A much better analogy is that of a partly open door which is in neither of the two extreme states – fully open and shut – nor swinging wildly between these, but is static in an intermediate position and having in some measure the properties of both the extreme states.

The greater the number of canonical structures, the more stable a species is; and the more alike these structure are, the more stable a species. The resonance hybrid is more stable than any of the canonical structures by an amount known as the delocalisation (or resonance) energy.

1.2 Isomerism and nomenclature

Isomerism is the phenomenon of two compounds of identical molecular formula having different molecular structures and is possible in alkanes containing more than three carbon atoms, when the possibility of straight or branched chain structures exists; e.g. C_4H_{10} can have two structures:

14 Basic principles: structure, nomenclature, stereochemistry and mechanism

[Structural formulae of two isomers of C₄H₁₀: n-butane and isobutane]

and C₅H₁₂ can have three:

[Structural formulae of three isomers of C₅H₁₂: n-pentane, isopentane, and neopentane]

With increasing values of n in the alkanes, C_nH_{2n+2}, the number of possible structures increases very rapidly, and $C_{20}H_{42}$ has approximately 4×10^5 isomers.

The diagrams above are known as the 'constitutional formulae' of the molecules represented. The constitution of a molecule describes the atomic sequence and nature of bonding between adjacent atoms in the molecule, but gives no information about the three-dimensional aspects of the molecular structure, such as shape or relative spatial arrangements of atoms or groups in a molecule.

Nomenclature

In order to simplify the naming of the vast number of organic compounds, a systematic form of nomenclature has been devised, in which the name of a compound is composed of syllables indicative of the functional groups present. All saturated hydrocarbons have the suffix '**-ane**' and in the unbranched ('normal') alkanes the previous syllable indicates the number of carbon atoms in the molecule:

Table 1.3

Formula	Name of unbranched alkane	Carbon skeleton of alkane
CH_4	Methane	C
C_2H_6	Ethane	C—C
C_3H_8	Propane	C—C—C
C_4H_{10}	Butane	C—C—C—C
C_5H_{12}	Pentane	C—C—C—C—C
C_6H_{14}	Hexane	C—C—C—C—C—C
C_7H_{16}	Heptane	C—C—C—C—C—C—C
C_8H_{18}	Octane	C—C—C—C—C—C—C—C
C_9H_{20}	Nonane	C—C—C—C—C—C—C—C—C
$C_{10}H_{22}$	Decane	C—C—C—C—C—C—C—C—C—C

The first four members of the series have old, non-systematic (trivial) names, and it will be found in the other groups of compounds that the simpler species retain their old trivial names, whilst the more complex molecules have been renamed systematically.

The branched chain alkanes can be regarded as derived from the unbranched alkanes by replacement of hydrogen atoms by 'substituent groups', each of which, consisting of an alkane minus a hydrogen atom, is known as an alkyl group; e.g. CH_3 methyl, C_9H_{19} nonyl. If the carbon atoms are numbered from one end, then the position as well as the structure of the substituent group can be described:

$$CH_3\underset{5}{-}CH_2\underset{4}{-}\underset{3}{\overset{\overset{\displaystyle CH_3}{|}}{CH}}-CH_2\underset{2}{-}CH_3\underset{1}{} \quad \text{3-Methylpentane}$$

For this purpose, the parent alkane is the one corresponding to the longest continuous chain of carbon atoms in the molecule, and the numbering of the carbon atoms of the parent alkane is started at the end that will give the lowest aggregate of numbers in the systematic name. The substituents are listed alphabetically and where two substituents are attached to the same atom, both are numbered, e.g. 2,2-dimethylbutane.

$$CH_3\underset{1}{-}CH_2\underset{2}{-}\underset{3}{\overset{\overset{\displaystyle CH_3}{|}}{CH}}-CH_2\underset{4}{-}\underset{5}{\overset{|}{CH}}-CH_3 \quad \begin{array}{c}\text{3,5-Dimethyloctane}\\[4pt]\textbf{not}\\ \text{4,6-dimethyloctane}\\ \text{3-methyl-5-propylhexane}\\ \text{2-ethyl-4-propylpentane}\end{array}$$
$$\underset{6\ 7\ 8}{CH_2CH_2CH_3}$$

$$CH_3-\overset{\overset{\displaystyle CH_3}{|}}{\underset{\underset{\displaystyle CH_3}{|}}{C}}-CH_2CH_3 \quad \text{2,2-Dimethylbutane}$$

The presence of a double bond in a molecule (alkenes) is indicated by the suffix '**-ene**', e.g. propene $CH_3CH=CH_2$, and this leads to structural isomerism:

$$\underset{1}{CH_3}-\underset{2}{\overset{\overset{\displaystyle CH_3}{|}}{CH}}-\underset{3}{CH}=\underset{4}{CH_2} \qquad \underset{1}{CH_3}-\underset{2}{\overset{\overset{\displaystyle CH_3}{|}}{C}}=\underset{3}{CH}-\underset{4}{CH_3}$$
$$\text{3-Methylbut-1-ene} \qquad\qquad \text{2-Methylbut-2-ene}$$

The presence of a triple bond in a molecule (alkynes) is indicated by the suffix '**-yne**'. In both alkenes and alkynes the parent structure is numbered so that the multiple bond has the lowest number possible. The parent structure is that which has the chain with the greatest number of multiple bonds.

$$CH_3CH_2\underset{2}{C}-\underset{3}{CH}=\underset{4}{CH}-\underset{5}{CH}=\underset{6}{\overset{\overset{\displaystyle|}{}}{C}}-\underset{7}{CH_3}$$
$$\underset{1}{\overset{\|}{CH_2}} \qquad\qquad\qquad CH_3$$

2-Ethyl-6-methylhepta-1,3,5-triene

In multifunctional compounds (see later chapters) one ending is added to indicate the most important functional group. The order of priority is:

CO₂H   > CONH₂   > CHO   > RCOR > NH₂   > OH
carboxy   carboxamido   formyl   oxo   amino   hydroxy
'-oic acid' '-oic amide'   '-al'   '-one'   '-amine'   '-ol'

Groups always named as substituents include halo (F, Cl, Br, I), alkoxy (OR) and nitro (NO_2).

1.3 Stereochemistry

1.3.1 Enantiomerism

So far, the simple compounds and reactions that have been described have been treated with little reference to the three-dimensional aspects of molecular structure. However, in more complex compounds the phenomenon of stereoisomerism and its consequences are of great significance, particularly in connection with many of the biologically important compounds.

Stereoisomerism is a form of isomerism in which compounds of the same constitutional formulae differ in the spatial arrangements of the functional groups. Stereoisomerism in simple molecules can be divided into **optical isomerism** (**enantiomerism**) and **geometrical isomerism** (*cis, trans* isomerism), but in complex molecules the distinction between these types may often be less clear-cut.

Optical isomerism arises from the possibility of having a three-dimensional structure which is not superimposable on its mirror image. Such structures are said to have the property of '**chirality**', or to be '**chiral**', and are characterised by the lack of a plane of symmetry (referring, of course, to the three-dimensional structure, and not to the representation on paper).

The simplest case of a chiral structure is that of an 'asymmetrically substituted carbon atom', i.e. a carbon atom bearing four, different, covalently bound substituents. Such a structure has two non-superimposable isomers, **enantiomers**, which are related as object to mirror image, e.g. (**15**) and (**16**).

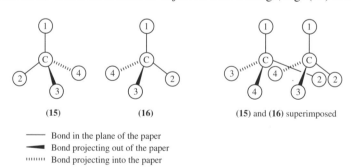

(15)   (16)   (15) and (16) superimposed

—— Bond in the plane of the paper
▬◄ Bond projecting out of the paper
⋯⋯ Bond projecting into the paper

However, if only two of the substituents are the same, this structure acquires a plane of symmetry, and becomes superimposable on its mirror image. In (**15**) and (**16**), if the substituents ③ and ④ are identical, then the structures will have planes of symmetry passing through ①, C and ②.

Stereochemistry

Numerous other types of chiral structure are known to chemists. However, in the compounds we shall consider in this and later chapters, the overwhelming majority of cases of optical isomerism will be concerned with structures in which the chirality is directly attributable to asymmetric substitution of a tetracovalent atom.

Polarised light

Light is an electromagnetic radiation composed of oscillating electric and magnetic fields, which are mutually perpendicular to each other and to the direction of propagation of the light beam (Figure 1.17). Normal light consists of multiple beams, which have random relative orientation of electric vectors. In plane-polarised light (usually referred to simply as 'polarised light') all the beams have their electric fields aligned parallel, with all the magnetic fields oscillating in the perpendicular plane.

Figure 1.17

Polarised light is unaffected by solutions of compounds whose structures have symmetry that precludes enantiomerism. However, it is found that if a beam of plane-polarised light is passed through a solution of one enantiomer of a compound with a chiral structure, the plane of polarisation of the light is rotated either clockwise or anticlockwise (Figure 1.18), and an equal and opposite rotation of the plane of polarisation occurs on passing through an equimolar solution of the other enantiomer.* The isomer whose solution rotates the plane of polarisation in a clockwise direction (with the observer facing the light source) is described as **dextrorotatory**; that which rotates the plane of polarisation in an anticlockwise direction is described as **laevorotatory**.

If solutions of enantiomeric compounds, which separately are dextro- and laevorotatory, are mixed to produce a solution containing equal concentrations of the two isomers, then this mixed solution is optically inactive. Such a mixture of enantiomers is known as a **racemic mixture**.

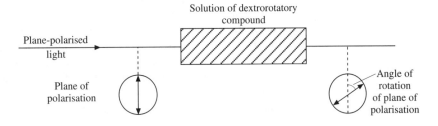

Figure 1.18

---

*These observations are made with an instrument known as a polarimeter. It is beyond the scope of this text to describe its method of operation, which, if required, should be sought in a textbook of physics.

The specific rotation (α) of a substance is defined as the angle (in degrees) through which the plane of polarisation is rotated when plane-polarised light is passed through a 10 cm length of a solution of concentration 1 g ml^{-1}. This is a characteristic property of chiral compounds (like the melting point), dextrorotatory compounds having their specific rotation described as positive, laevorotatory compounds being described as negative. Thus, if one isomer of an enantiomeric pair has a specific rotation of +150°, the other (laevorotatory) isomer will have a specific rotation of −150°. The symbols (+) and (−) are used in nomenclature to distinguish between two enantiomers; thus the two isomers of 2-bromobutane, $CH_3CH_2CHBrCH_3$, can be described as (+)-2-bromobutane (i.e. dextrorotatory) and (−)-2-bromobutane (laevorotatory), and the racemic mixture of these as (±)-2-bromobutane.*

Although enantiomers have markedly different effects on polarised light, all the other physical properties of the separate enantiomers, and all those chemical properties which do not involve other chiral molecules, are identical. Thus two enantiomeric carboxylic acids (e.g. (+)- and (−)-$C_6H_5CH(C_2H_5)CO_2H$) will have the same melting and boiling points, and the same refractive index, density, solubility and viscosity. The pKs will be identical, and they will form esters with, say, methanol or ethanol, whose properties, other than optical rotation, will likewise be identical. However, they will react at different rates with, say, (+)-butan-2-ol, and the two esters will not now be identical (p. 21).

The synthesis of a compound with a chiral structure usually produces both enantiomers in equal proportions (i.e. a racemic product), e.g.

and the close similarity of the properties of the two enantiomers makes their separation difficult. No simple physical separation by techniques such as fractional distillation or recrystallisation is possible. However, the following methods have been employed to **resolve** (i.e. separate) racemic mixtures.

Mechanical separation. If a solution of a racemic mixture is allowed to crystallise, two types of product may be obtained. Either one type of crystal (a **racemate**) will separate, in which case the crystal lattice is built up of equal numbers of molecules of each of the enantiomers, or the solution will deposit a mixture of two types of crystal, one composed solely of the (+) enantiomer, the second containing solely the (−) enantiomer. In the latter case, if the two types of crystal can be distinguished, and if the individual crystals are big enough, then they can be separated by hand-picking. This very laborious and unsatisfactory method, which is rarely possible, is now only of historical interest, being the way in which Pasteur first resolved sodium ammonium (±)-tartrate. Many racemic mixtures crystallise as racemates and cannot,

---

*The archaic equivalents *d*, *l* and *dl* are sometimes encountered, and refer solely to the direction of rotation of the plane of polarised light.

therefore, be resolved by this method. It may be noted in passing that the racemate, having a different crystal structure, may have a widely different melting point and solubility from those of the separate enantiomers, and cases are known where mixing the saturated solutions of the enantiomers produces a precipitate of the less soluble racemate. These differences correspond to the relative ease of packing 'right-handed' and 'left-handed' molecules alternately into the crystal lattice, compared with forming a lattice from right- or left-handed molecules alone.

Resolution via diastereoisomeric compounds. If a racemic mixture of the two enantiomers of a chiral reagent is allowed to react with an optically inactive compound, the enantiomers will react separately, producing two enantiomeric products. This can be illustrated by the analogy of two pieces of rod with left- and right-hand screw threads (representing the enantiomers) being joined to another piece of rod which has no such chiral pattern (Figure 1.19a).

However, if the racemic mixture reacts with *one enantiomer of a second chiral reagent*, then the two products are no longer enantiomeric (i.e. related as object and mirror image), although they are stereoisomeric. Stereoisomers that are not mirror images are known as **diastereoisomers**, and diastereoisomers have physical properties that may differ widely, permitting separation by techniques such as fractional recrystallisation, distillation or chromatography.

A familiar illustration of the formation of diastereoisomers is in the insertion of left and right hands (which have a mirror-image relationship) into a right-hand glove, resulting in a situation in which (left hand + right-hand glove) is not the mirror image of (right hand + right-hand glove). It should also be noted that the latter process is much easier than the former, corresponding to the different rates at which enantiomeric compounds react with a chiral reagent.

The formation of diastereoisomers is the basis of the best method of resolving racemic mixtures. The racemic mixture is allowed to react with an 'optically pure' chiral reagent (i.e. only one enantiomer present), and the resultant mixture of diastereoisomeric products is separated by some suitable physical method. Afterwards, the original enantiomers are regenerated separately by reversal of the initial reaction. Thus a racemic mixture of an acid may be converted, by reaction with an optically pure chiral base, into crystalline, diastereoisomeric salts. After separation of the two salts by fractional recrystallisation, treatment with mineral acid liberates the enantiomeric acids:

(±) acid + (+) base → (+) acid (+) base + (−) acid (+) base
(+) acid (+) base + HCl → (+) base hydrochloride + (+) acid
(−) acid (+) base + HCl → (+) base hydrochloride + (−) acid

This method of resolution is indirectly related to the following one in that nature is the only plentiful source of optically pure, chiral reagents. Many of the resolutions performed utilise complex natural bases (e.g. quinine, cinchonine or strychnine) having chiral structures of which only one enantiomer occurs naturally.

20 Basic principles: structure, nomenclature, stereochemistry and mechanism

Compounds that do not react with these bases can often be converted into derivatives that form salts. Thus a racemic alcohol may be converted into the half ester of a dibasic acid such as phthalic acid (benzene-1,2-dicarboxylic acid) and the resulting racemic carboxylic acid resolved via the formation of diastereoisomeric salts. After resolution, the chiral esters can be hydrolysed to regenerate the enantiomeric alcohols (Figure 1.19b).

In principle, resolution of a racemic mixture should be possible by chromatography using a chiral stationary phase, since adsorption of the racemic solute onto the chiral stationary phase will produce diastereoisomeric

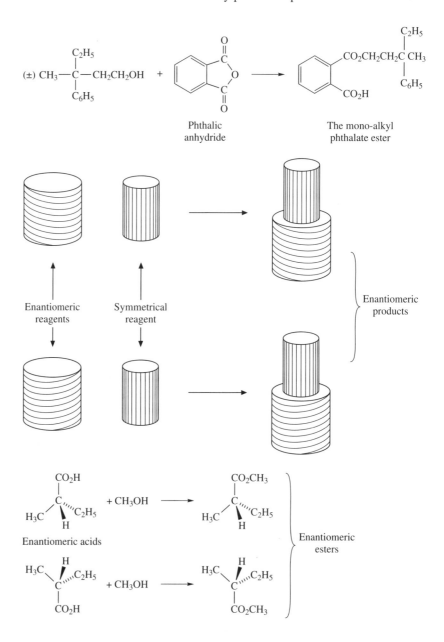

Figure 1.19a

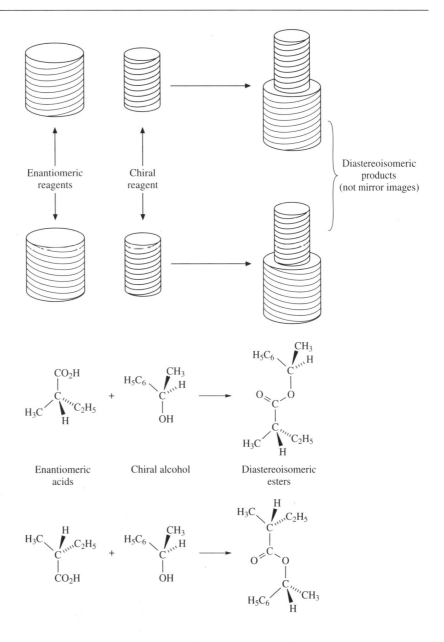

Figure 1.19b

adsorbent–solute combinations. One enantiomer of the racemic solute (the less tightly bound) should be eluted more rapidly than the other. In practice, this method has achieved varying degrees of success, lactose, starch, cellulose and optically active quartz being examples of the chiral adsorbents employed. Paper chromatography has also been used to achieve some resolutions either using the cellulose of the paper as the chiral adsorbent or by impregnation of the paper with some other chiral reagent. Thus 2-aminophenylethanoic acid, $C_6H_5CHNH_2CO_2H$, has been resolved by chromatography on paper impregnated with (+)-camphor-10-sulphonic acid.

(+)-Camphor-10-sulphonic acid

If a racemic mixture of one chiral reagent (A) reacts with a racemic mixture of a second chiral reagent (B), then four products are obtained:

$$(\pm)A + (\pm)B \longrightarrow \begin{cases} (+)A\,(+)B & (-)A\,(-)B \\ (+)A\,(-)B & (-)A\,(+)B \end{cases}$$

As the products are set out above, horizontal pairs are enantiomeric, and vertical or diagonal pairs are diastereoisomeric.

Biological resolution of racemic mixtures. If a racemic mixture can be fed to a living organism, then it is often found that one enantiomer is preferentially metabolised. If this is so, then the unwanted isomer can sometimes be recovered. When a racemic mixture of mevalonic acid (3,5-dihydroxy-3-methylpentanoic acid) is fed to rats, one optical isomer is totally absorbed and almost all the other is excreted in the urine, from which it can be recovered. Similarly, moulds or other micro-organisms will utilise one enantiomer of a racemic mixture in the culture medium. However, this method suffers from several disadvantages: compounds may be poisonous or not assimilated at all; and even if the method works, one of the enantiomers is always lost.

Mevalonic acid

(metabolised)   (excreted)

This method of resolution is really another example of resolution via formation of diastereoisomers. Reactions occurring in living systems are controlled by protein catalysts (enzymes) which are themselves chiral. The ability of an organism to metabolise a substance depends upon the presence of enzymes which will adsorb the molecules prior to catalysing their chemical transformation as part of the process of digestion. The initial formation of the enzyme–substrate complex is just another case of the interaction of a single enantiomer of a chiral reagent (the enzyme) with a racemic compound, and that enantiomer of a racemic substrate which most readily combines with the enzyme will be preferentially metabolised.

This method of resolution has been adapted to a more convenient form by the use of purified enzymes to overcome the numerous practical disadvantages of employing living systems. Thus, a racemic amine may be resolved by conversion into its *N*-ethanoyl derivative followed by enzymic hydrolysis of the racemic amide. By appropriate choice of enzyme and conditions, one of the enantiomers of the amide can be hydrolysed selectively, leaving a mixture of amine and amide which is easily separated by standard chemical means, e.g.:

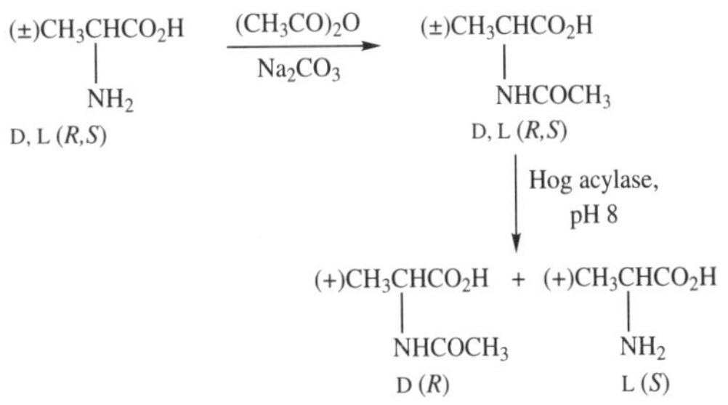

$$(\pm)CH_3CHCO_2H \xrightarrow[Na_2CO_3]{(CH_3CO)_2O} (\pm)CH_3CHCO_2H$$
$$\underset{D,L\,(R,S)}{|\atop NH_2} \qquad \underset{D,L\,(R,S)}{|\atop NHCOCH_3}$$

$$\xrightarrow[pH\,8]{Hog\ acylase}$$

$$(+)CH_3CHCO_2H \; + \; (+)CH_3CHCO_2H$$
$$\underset{D\,(R)}{|\atop NHCOCH_3} \qquad \underset{L\,(S)}{|\atop NH_2}$$

The designation of chirality by D and L, and the representation of absolute configuration

As can be seen from the previous example, in which the (−)-enantiomer of the amide is hydrolysed to the (+)-amine, *there is no simple relationship between the direction of rotation of polarised light and the molecular configuration.* For reasons that will become apparent later, the dextrorotatory isomer of glyceraldehyde was taken as the standard compound for carbohydrates, arbitrarily assigned the configuration shown below, and named D-glyceraldehyde. Its enantiomer was named L-glyceraldehyde. In these names D and L

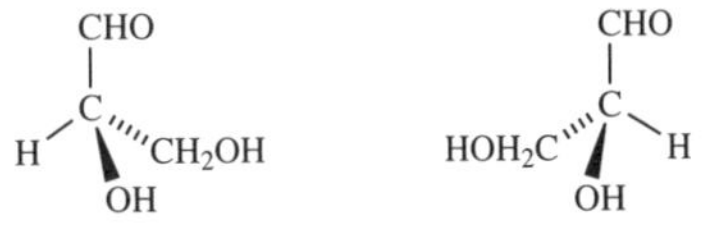

D-(+)-Glyceraldehyde   L(−)-Glyceraldehyde

refer to configuration only, and are not related to optical rotation. (They should not be confused with *d* and *l*). All other sugars (or their derivatives) are then related, by a series of chemical transformations which effectively convert the glyceraldehyde to the sugar, back to D- or L-glyceraldehyde.

When recently it became possible to determine the absolute configuration of enantiomers (by a method employing diffraction of X-rays by crystals in a way which permitted the absolute spatial disposition of atoms to be determined) it was found that, purely by chance, the correct assignment of configuration had been made. The projection formulae above, therefore, represent the true absolute configurations of the enantiomers. (See p. 161 for the use of D and L in amino acids.)

The Fischer projection

While it is possible to draw two-dimensional diagrams which are reasonably clear perspective diagrams of three-dimensional structures having one or two chiral centres, this is not so for more complex molecules, and a stylised representation known as the Fischer projection is used for this purpose. The two enantiomers of the species C①②③④ are represented by:

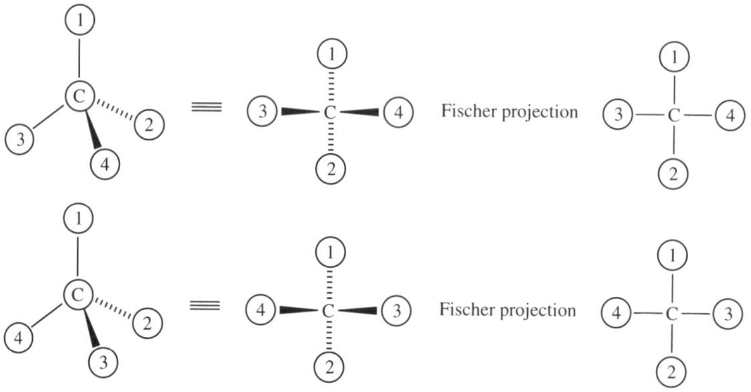

the convention being that the horizontal bonds in the Fischer projection project forward from the plane of the paper, and the vertical bonds project backwards into the paper. Thus, in the Fischer projection of D-glyceraldehyde, the hydroxyl group is to the right of the carbon chain, when this is represented with the aldehyde group at the top, and vice versa for L-glyceraldehyde.

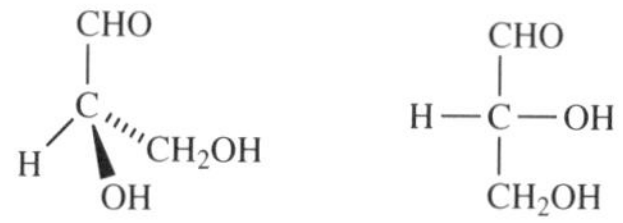

D-(+)-Glyceraldehyde

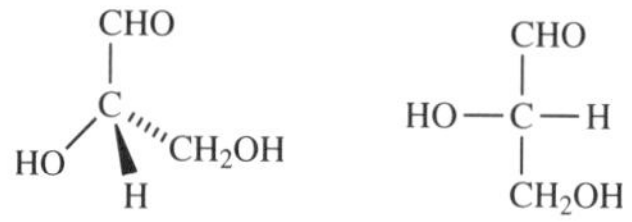

L-(−)-Glyceraldehyde

When using the Fischer projection, care must be taken about rotation of the projection formulae, e.g.

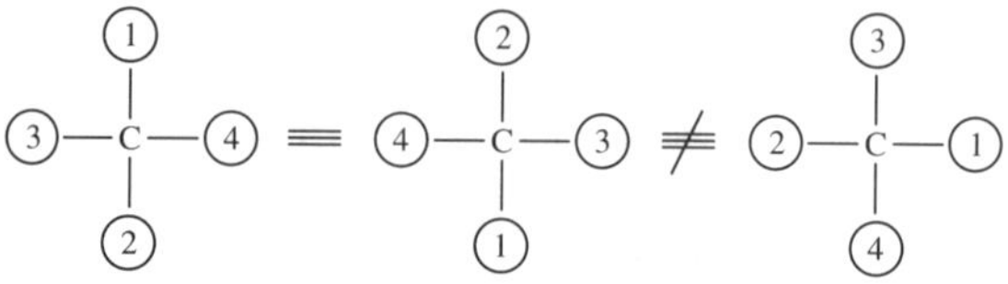

i.e. rotation of the projection formulae through 180° produces the projection of the same configuration and rotation through 90° produces the projection of the enantiomer, since exchange of the vertical and horizontal bonds in the projection is equivalent to a mirror-image inversion of the three-dimensional structure.

1.3.2 Compounds with more than one chiral centre

Since there are two possible configurations for an asymmetrically substituted carbon atom, a structure containing n such asymmetric centres should have 2^n stereoisomers. This is the maximum number, but in some cases the number of possible stereoisomers is less than this.

If we consider a compound with two chiral centres, in which the asymmetrically substituted atoms have quite different series of groups attached, then the expected stereoisomers are obtained, corresponding to structures (17)–(20):

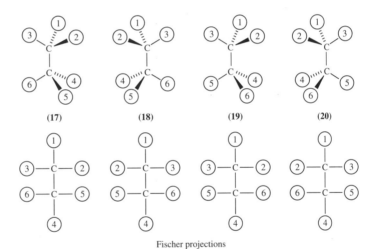

Fischer projections

In these four structures, (17) and (18) are enantiomers, as are (19) and (20), and in general with n chiral centres there will be 2^{n-1} pairs of enantiomers.

If, however, we consider the case of a compound with two chiral centres, each bearing identical substituent groups, only three stereoisomers are possible. This can be seen by supposing that in structures (17)–(20) group ① is identical to ④, ② to ⑤, and ③ to ⑥, when the four stereoisomers become (21)–(24). Of these (21) and (22) are identical, superimposable, structures since they have a plane of symmetry perpendicular to the C—C bond. There are, therefore, only three stereoisomers possible in this case, of which two, (23) and (24), are enantiomeric and therefore optically active, and one, represented by (21) or (22), is optically inactive since it is superimposable upon its mirror image. Such an optically inactive stereoisomer is designated by the prefix 'meso'.

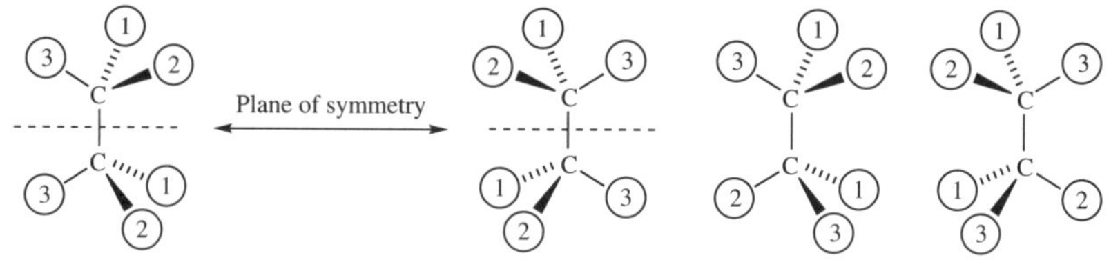

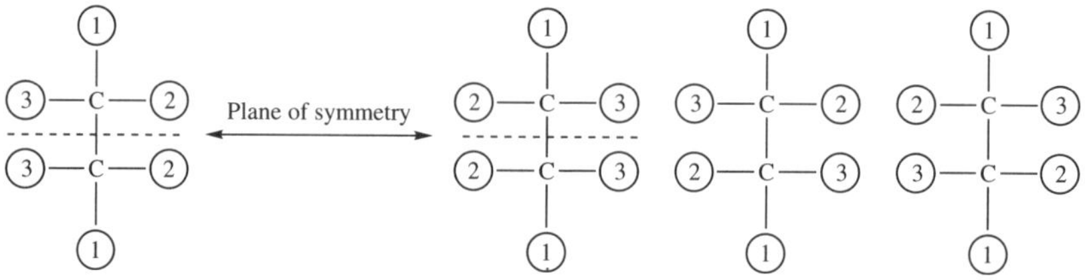

Tartaric acid, HO$_2$CCH(OH)CH(OH)CO$_2$H, is a well-known example of this phenomenon. Three stereoisomers are known, whose structures and Fischer projections are shown below. Whilst (+)- and (−)-tartaric acid, being enantiomers, will have identical physical properties, this is not so for the *meso*-stereoisomer, which will have different physical properties.

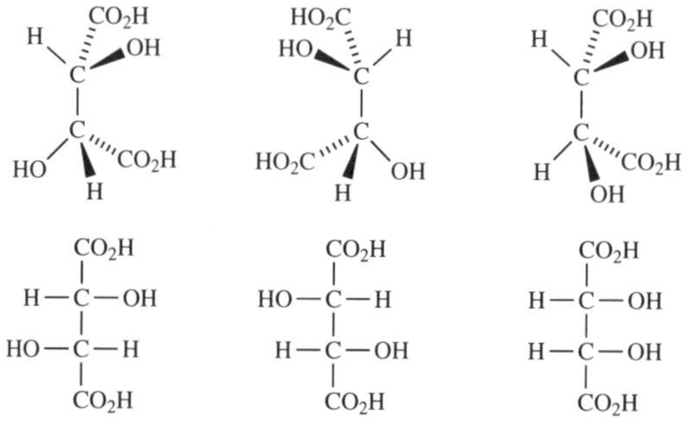

(+)-Tartaric acid   (−)-Tartaric acid   *meso*-Tartaric acid

If, in a chemical reaction, the functional groups in a *meso*-compound react in such a way as to destroy the molecular symmetry, then two enantiomeric products will be obtained. Thus *meso*-tartaric acid forms two optically active, enantiomeric monomethyl esters, CH$_3$O$_2$CCH(OH)CH(OH)CO$_2$H, both of which, on further esterification, form the same optically inactive dimethyl ester, CH$_3$O$_2$CCH(OH)CH(OH)CO$_2$CH$_3$.

1.3.3 Designation of chirality by *R* and *S* (Cahn–Ingold–Prelog rules)

The use of D- and L- and the Fischer projection to describe absolute configuration has a number of disadvantages arising from the necessity of relating compounds structurally to glyceraldehyde. In some cases the same chiral structure can equally readily be related to either D- or L-glyceraldehyde depending upon the hypothetical chemical transformations chosen. Tartaric acid is such a case, the scheme below showing how either (+)- or (−)-tartaric acid can be derived from D-glyceraldehyde with its configuration preserved throughout the sequence. (NB The chemical transformations are hypothetical, and in each scheme the box encloses the chiral centre whose configuration is derived from D-glyceraldehyde.)

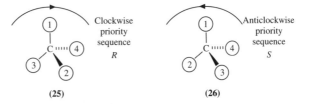

A more systematic method of denoting absolute configuration has now been proposed. If a carbon atom is bonded to four different groups ①, ②, ③ and ④, and we assign an arbitrary sequence of priority 1, 2, 3, 4, then the two configurations at the asymmetrically substituted atom can be distinguished by inspection of the tetrahedron from the side remote from the group of lowest priority (i.e. ④), when the other three groups will be seen in a clockwise (25) or anticlockwise (26) sequence of decreasing priority. These configurations are described by R (from *rectus*) and S (from *sinister*) respectively.

The arbitrary priority rules adopted are that the highest priority goes to the group joined to the chiral centre by the atom of highest atomic number (e.g. I, Br, Cl, SH, F, OH, NH_2, CH_3, H in this order of decreasing priority). Hydrogen, therefore, always has the lowest priority. If the chiral centre is attached to two isotopes of the same atom then the priority goes to the heavier isotope (e.g. T > D > H and $^{14}C > ^{13}C > ^{12}C$). If two groups have identical atoms joined to the chiral centre then a distinction between these groups is sought by comparison of the atomic numbers of the atoms one stage further removed (e.g. $CH_2Cl > CH_2OH > CH_2CH_3 > CH_3$). When, at this point, more than one atom of highest atomic number has to be considered, then priority goes to the group with most such atoms in second place (e.g. $C(CH_3)_3 > CH(CH_3)_2 > CH_2CH_3$), and additionally, doubly bonded atoms count twice and triply bonded atoms count three times.

If it is still not possible to assign a priority sequence, the atomic numbers of the next set of atoms are taken into account. It is important to realise that the process of priority assignment stops immediately a distinction can be made, irrespective of the atomic numbers and frequency of atoms which might be taken into account at the next step. Thus $CH(CH_3)_2 > CH_2CBr_3$ since the former has two carbon atoms in second place as opposed to one in the latter group.

Although this system of nomenclature is almost invariably employed by chemists, it has not yet entirely displaced the older D/L designation for simple biological molecules. Throughout the remainder of this book absolute configurations will be described in terms of the Fischer projection using both the D/L and *R/S* nomenclatures where this is possible, since the D/L system is likely to persist in biochemical usage, notwithstanding its disadvantages.

1.3.4 Asymmetric substitution of atoms other than carbon

Although optical isomerism has, so far, been described solely in terms of carbon compounds, any tetrahedrally substituted atom is potentially a source of chirality.

Ammonia and amines have a distorted tetrahedral structure (p. 8) with a lone pair of electrons acting as the fourth substituent. However, ammonia and amines invert extremely rapidly, so that although chirality may be present in the structures, enantiomers can never be separated. Quaternary ammonium cations and tertiary amine *N*-oxides, on the contrary, have stable configurations, and enantiomers can be resolved.

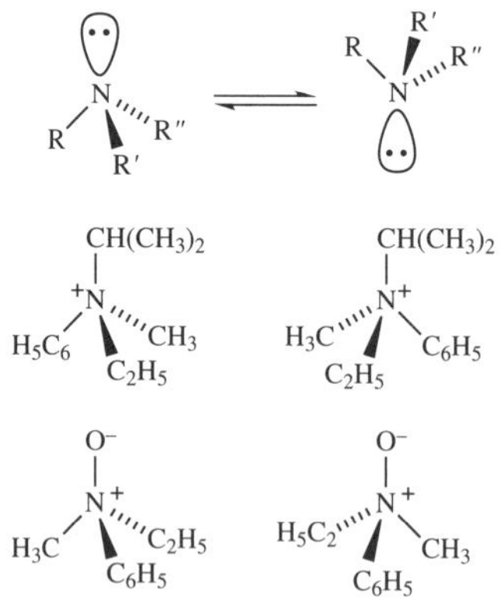

Sulphonium cations have tetrahedral configurations, which, unlike amines, do not readily invert, and sulphoxides (p. 74) also have stable tetrahedral configurations. Examples of both types of compound have been resolved into enantiomers, as have asymmetric phosphorus compounds (see p. 75).

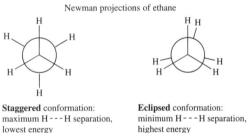

A sulphoxide

1.3.5 Conformational isomerism

Conformational isomerism in open-chain molecules

There is much evidence to suggest that in simple, open-chain molecules rotation about single bonds can and does occur readily. Thus in ethane the two methyl groups rotate independently about the central σ bond, whose electron cloud is axially symmetrical about the line joining the carbon nuclei (p. 6), leading to an infinite number of **conformations** for the ethane molecule (different conformations of a molecular structure can be interconverted by rotation about bonds, i.e. without the breaking and re-formation of chemical bonds; cf. 'configuration', p. 33). However, although this rotation occurs very easily, it is not wholly unrestricted since the different conformers of the ethane molecule vary slightly in energy on account of the differing separation between hydrogen nuclei attached to adjacent carbon atoms and the consequent variation in interaction between the electron clouds of the C—H bonds.

Newman projections of ethane

Staggered conformation: maximum H - - - H separation, lowest energy

Eclipsed conformation: minimum H - - - H separation, highest energy

These variations and, in particular, the conformations of maximum and minimum energy, are best illustrated by the **Newman projections**, which are visualisations of the molecule viewed along the bond about which rotation is being considered. In the foreground of these projections are three bonds linked to the near carbon atom, behind which the circle represents the electron cloud of the σ bond. The three bonds to the remote carbon atom project from the rear of the electron cloud. The relative orientations of C—H bonds about the C—C bond are easily seen from this projection and the highest energy '**eclipsed**' and lowest energy '**staggered**' conformations are readily discerned and easily portrayed.

The type of interaction that gives rise to very small energy differences between the conformational extremes for ethane causes much greater energy differences where there are larger substituents on the carbon atoms. Thus,

30 Basic principles: structure, nomenclature, stereochemistry and mechanism

1,2-dibromoethane, CH₂BrCH₂Br, has three staggered conformations, of which the one with the maximum separation between the bulky bromine atoms has the lowest energy, and three eclipsed conformations of which that with the two C—Br bonds eclipsed, giving the closest approach of the large bromine atoms, has the highest energy. At any moment in a sample of this compound a substantial proportion of the molecules will be in conformations closest to that of the lowest energy and only a minute fraction will be in the highest energy eclipsed conformation. Similar considerations apply to all singly bonded systems where free rotation can occur, and the conformational preferences described for dibromoethane apply equally well to conformers arising from rotation about the central bond of butane. Where the chemical properties of these or other compounds depend upon the precise shape of the molecule, conformational preferences of these types may be very important in determining the course or fate of reactions.

Conformational isomerism in cyclic systems

The conformational features of open-chain molecules described above have significant effects upon the shapes of small, cyclic molecules and their derivatives, which will be considered briefly here. The lower cycloalkanes serve as model compounds for saturated three-, four-, five- and six-membered rings in general. In these structures free rotation is not possible about the single bonds that form the ring, but in four-membered and larger rings there is a restricted degree of rotational movement possible.

Cyclopropane has a planar ring of carbon atoms (since three points define a plane), and if the valency bonds between carbon atoms are considered to run linearly between the carbon nuclei then the C—C—C bond angle = 60°, which is very considerably less than the angle of 109.5° found in tetrahedrally

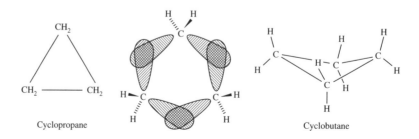

substituted carbon. Alternatively, the small ring can be considered to be formed by less than optimum overlap of orbitals, as shown above. It is found that cyclopropane is a very reactive hydrocarbon (catalytic hydrogenation converts it into propane) and is said to have a 'strained' ring. Around this planar ring adjacent C—H bonds are inevitably eclipsed. Cyclobutane, which if planar would have $\angle$C—C—C = 90°, is slightly buckled, with a dihedral angle of approximately 170°. Although this buckling reduces the angle $\angle$C—C—C to slightly less than 90°, increasing the ring strain with respect to the planar structure, it reduces the unfavourable eclipsing interactions between adjacent C—H bonds in the planar conformation. Small rotations about the ring bonds will interconvert the two possible buckled structures. Cyclobutane is less strained than cyclopropane and is therefore less reactive, being inert to hydrogenation (like all the larger cycloalkanes).

A planar ring of five carbon atoms would have $\angle$C—C—C = 108°, which is very close to the tetrahedral angle, but to relieve the associated eclipsing C—H interactions cyclopentane adopts a slightly buckled structure, rather like an open envelope with the flap raised. Here too, limited rotation about ring bonds interconverts a series of geometrically identical conformations in which different carbon atoms occupy the raised position.

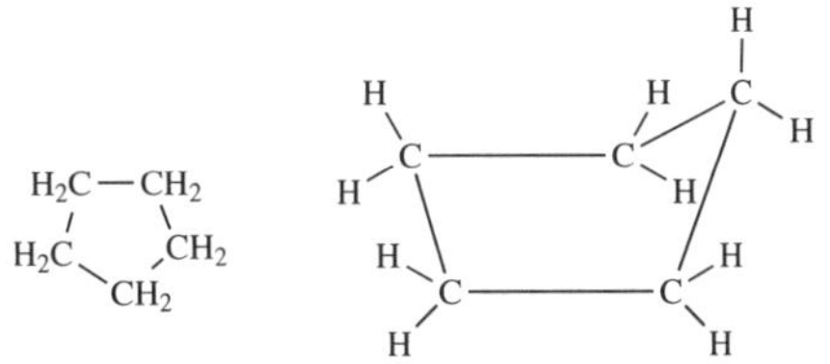

Cyclopentane

If cyclohexane had a regular planar ring of carbon atoms, $\angle$C—C—C would be 120°, which is considerably greater than the preferred tetrahedral angle. Since puckering of a planar ring results in a decrease in angles, cyclohexane adopts buckled structures in which $\angle$C—C—C = 109.5°. Two such

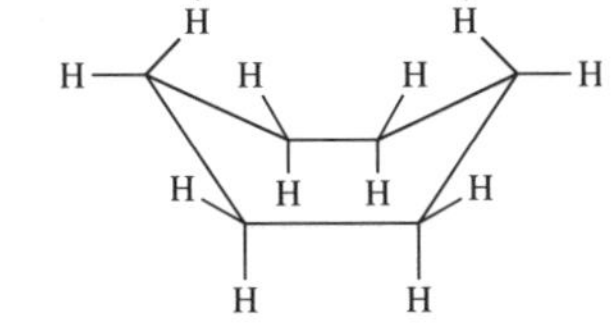

Cyclohexane (boat conformation)

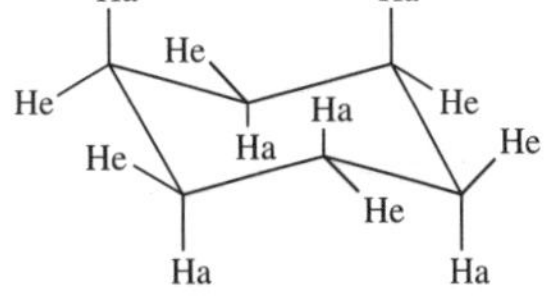

Cyclohexane (chair conformation)

C—Ha axial bond
C—He equatorial bond

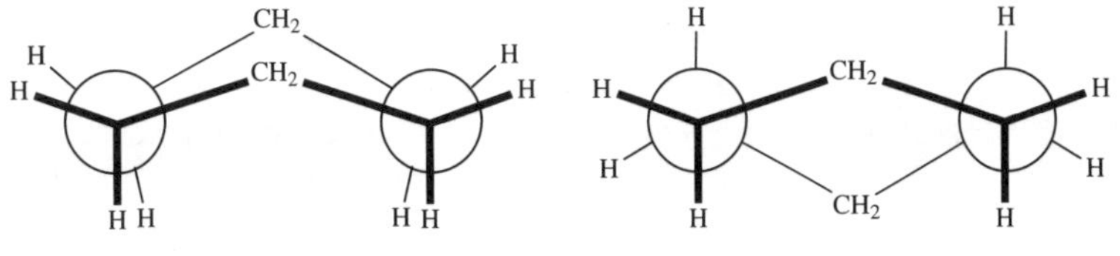

Newman projections

structures are possible, known as the 'boat' and 'chair' conformations. The boat form is not important in practice since in this structure two of the hydrogen atoms (marked with asterisks) approach so closely that their electron clouds overlap, causing strong repulsion, and neighbouring C—H bonds along the sides of the boat are eclipsed. Together these give the boat conformation of cyclohexane a higher energy than the alternative chair conformation, such that at room temperature only one molecule of cyclohexane in a thousand is in the boat conformation. In the chair conformation these 'non-bonded interactions' are minimised, with adjacent CH$_2$ groups being staggered. However, in the chair form of cyclohexane two types of C—H bond orientation can be discerned, with six C—H bonds lying parallel to the axis of the ring ('axial' orientation) and the other six directed out to the sides of the ring ('equatorial' orientation). As a result there could, in principle, be two isomeric monosubstituted cyclohexanes, e.g. chlorocyclohexane, with the substituent occupying an axial orientation in one case and an equatorial orientation in the other. In fact both species exist and are rapidly interconverted by limited rotational movements around the bonds of the ring, which invert one chair conformation to form the other via the intermediate formation of a boat conformation. Usually, the conformation in which a substituent group is in an axial position is energetically disfavoured with respect to the equatorial conformer. In the axial orientation, close approach to the axial hydrogen atom (or other groups) on the same side of the ring causes repulsive overlap of the electron clouds. In the equatorial orientation the corresponding interactions with neighbouring, but more distant, equatorial hydrogen atoms are less significant. In general, therefore, substituted cyclohexane molecules tend to adopt the chair conformation in which a substituent lies (or, for multiply substituted derivatives, most substituents lie) in an equatorial orientation. If a very large substituent group such as *t*-butyl, C(CH$_3$)$_3$, is attached to a six-membered ring, this effectively locks the cyclohexane ring in that chair conformation in which the bulky group is equatorial.

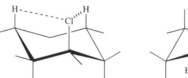

Strong non-bonded interaction   Weak non-bonded interaction

Since the preferred bond angles for trivalent nitrogen and divalent oxygen are approximately the same as that for sp^3 hybridised carbon (p. 9), the same considerations apply to six-membered rings containing nitrogen and oxygen atoms. Piperidine and tetrahydropyran adopt chair-shaped conformations similar to that of cyclohexane, in which C—H bonds are replaced by lone pairs at the heteroatom sites.

Larger saturated rings containing seven or more carbon atoms are known. All have buckled structures in which all ∠C—C—C = 109.5°, but the flexibility of the rings makes their stereochemistry much more difficult to describe in general terms.

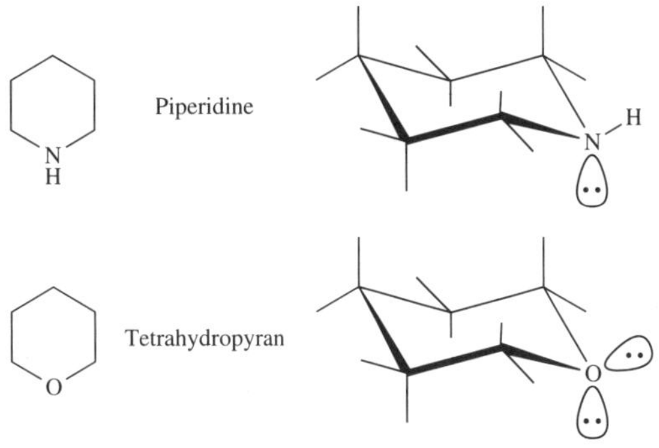

1.3.6 *Cis–trans* (geometrical) isomerism

The free rotation about σ bonds described above, which is a characteristic feature of open-chain, singly bonded molecules, does not readily occur about double bonds. In a molecule where two groups are joined by a double bond, the σ bond has the customary axial symmetry, but the π bond requires the p orbitals on adjacent atoms to be aligned parallel in space to achieve maximum overlap (p. 10). Any independent rotation of these groups will shift the p orbitals from this optimum alignment, thereby reducing overlap, and for rotation to occur freely the π bond must be broken. Since the breaking of a chemical bond requires the provision of a great deal of energy, molecules of the type (A)(B)C=C(A)(B) can exist in two different **configurations**:

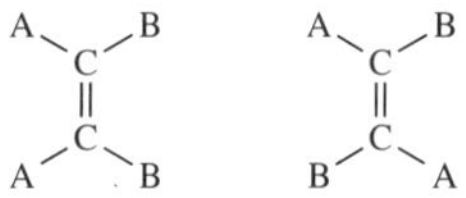

which can often be isolated or separated as distinct compounds with different chemical and physical properties. These stereoisomers are often distinguished by use of the prefixes *cis-* and *trans-*, *cis-* being employed to denote the compound in which similar groups are on the same side of the plane of the π bond, while *trans-* describes the isomer in which similar groups are on opposite sides. This convention is unsuitable for describing compounds in which four widely differing groups are attached to the doubly bonded atoms and the use of *cis* and *trans* in systematic nomenclature to designate configuration in these types of molecule has been superseded by the *E/Z* notation (p. 34) although they are still often employed in trivial (i.e. non-systematic) names.

cis-But-2-ene   *trans*-1,2-Dibromoethene

Cis–trans isomers (sometimes referred to as *cis–trans* diastereoisomers, since they are stereoisomers which are not enantiomers; p. 19) can, and do, differ in chemical and physical properties. Since in the solid state the crystal lattices of the isomers are built of quite different-shaped units, it is not surprising that differences occur in solubility and melting point, which reflect the differing stabilities of the crystal lattices. In simple cases it is frequently the isomer with similar groups *trans* which has the higher melting point and lower solubility since the more symmetrical *trans*-molecule packs better into a three-dimensional structure than the less symmetrical molecule with similar groups *cis*. In view of these differences, separation is usually possible by physical methods such as fractional distillation, fractional recrystallisation and chromatography. Simple *cis–trans* isomers have no effect on polarised light, though in complex molecules *cis–trans* and optical isomerism may occur simultaneously.

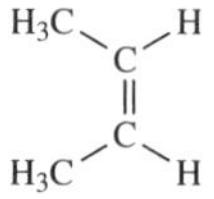

cis-But-2-ene
melting point = −139°C
boiling point = +3.7°C

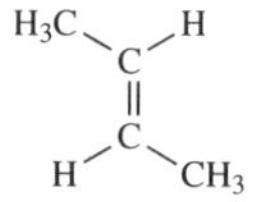

Trans-But-2-ene
melting point = −106°C
boiling point = +0.96°C

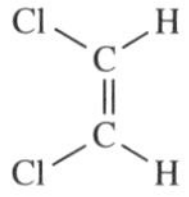

cis-1,2-Dichloroethene
melting point = −80.5°C
boiling point = +60.2°C

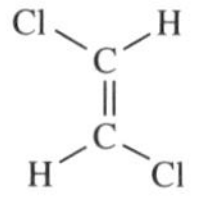

trans-1,2-Dichloroethene
melting point = −50.5°C
boiling point = +48.3°C

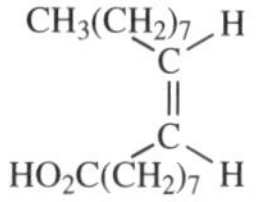

Oleic acid
(*cis*-Octadec-9-enoic acid)
melting point = 16°C

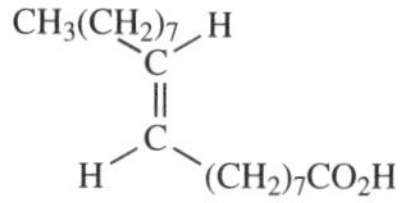

Elaidic acid
(*trans*-Octadec-9-enoic acid)
melting point = 54°C

Molecules with double bonds: designation of configuration by *E* and *Z*

The confusion about the precise meaning of the prefixes *cis* and *trans* and the difficulty associated with their usage when the four substituents on a doubly bonded system have no obvious relationship (e.g. $CH_3CH=CClBr$) have led to this method for designating configuration being replaced by one based on the Cahn–Ingold–Prelog rules for designating chirality (p. 27). Using these rules, a priority sequence can be assigned to the substituents at each end of the doubly bonded system, leading to two possibilities:

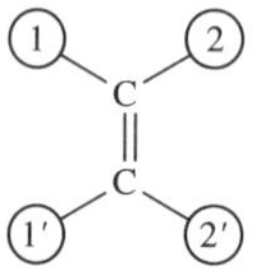

Z

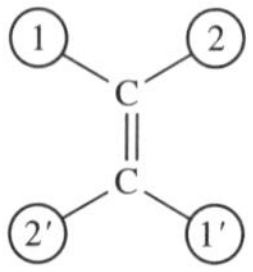

E

The configuration in which the high-priority substituents are on the same side is described as *Z* (German *zusammen* – together) and the alternative configuration is described as *E* (German *entgegen* – opposed). This system can be used for *cis–trans* isomers involving other than C=C bonds, and in these cases, where one of the doubly bonded atoms bears a lone-pair of electrons, the lone-pair has lower priority even than H, e.g.

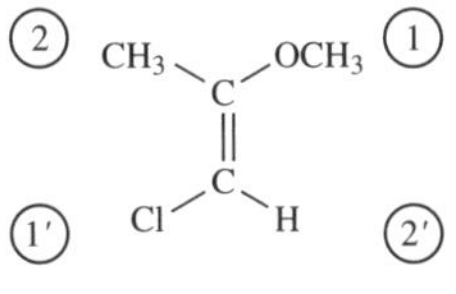

E-1-Chloro-2-methoxypropene

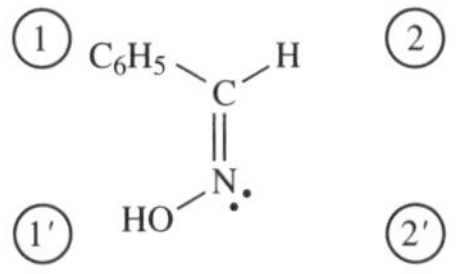

Z-Benzaldoxime

Cis–trans isomerism in cyclic compounds

The phenomenon of restricted rotation is not confined to molecules containing double bonds. In saturated cyclic molecules the difficulty of distortion of bond lengths and bond angles restricts the extent to which rotation may occur about the σ bonds making up the ring, and *cis–trans* isomers can be formed which differ in the orientation of substituents with respect to the plane of the ring. This is illustrated for 2,2,4,4-tetramethylcyclobutane-1,3-diol. The physical and chemical properties of these stereoisomeric compounds differ, but both molecules have a plane of symmetry (the plane of the paper) and are therefore optically inactive.

trans-2,2,4,4-Tetramethyl-cyclobutane-1,3-diol
melting point = 148°C

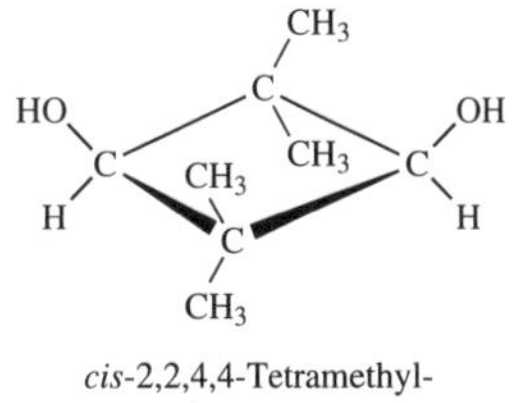

cis-2,2,4,4-Tetramethyl-cyclobutane-1,3-diol
melting point = 163°C

Similarly, with cyclohexane, although the larger ring gives greater flexibility to the molecule and interconversion between alternative chair conformations of each isomer is possible, a *cis*-disubstituted cyclohexane cannot be converted into its *trans* isomer by rotational processes alone, e.g. for 1,4-dibromocyclohexanes:

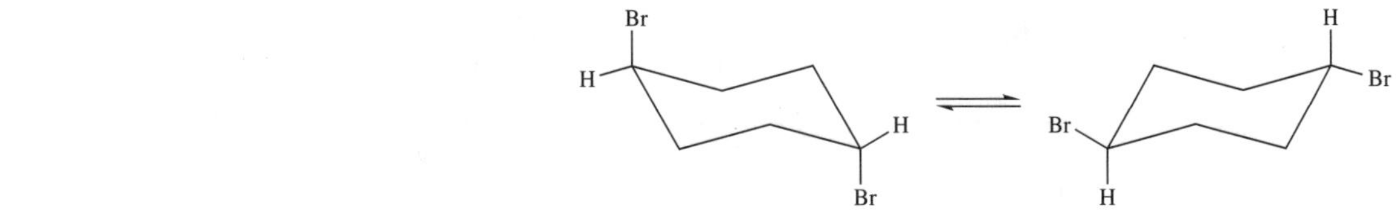

trans-1,4-Dibromocyclohexane

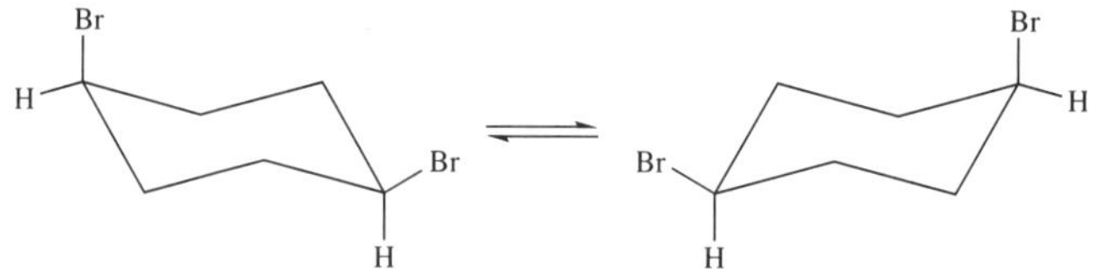

cis-1,4-Dibromocyclohexane

In these examples too, both stereoisomers have a plane of symmetry, precluding enantiomerism.

A more complex situation exists in other cases, such as the cyclopropane-1,3-dicarboxylic acids. The *cis* isomer has a plane of symmetry (perpendicular to the ring, passing through C-3 and bisecting the bond between C-1 and C-2) and is therefore optically inactive. The *trans* isomer has no plane of symmetry and therefore exists in enantiomeric forms. While the enantiomeric *trans*-dicarboxylic acids will have identical chemical and physical properties (except

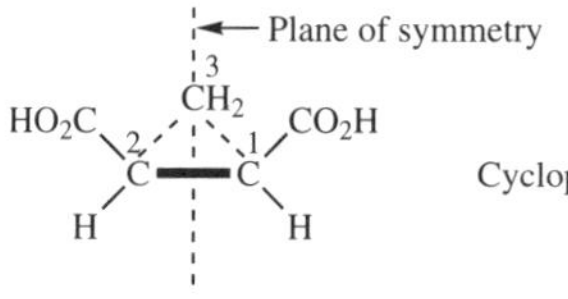

Cyclopropane-1(R), 2(S)-dicarboxylic acid

cis-Cyclopropane-1,2-dicarboxylic acid
melting point = 139°C; pK_1 3.40

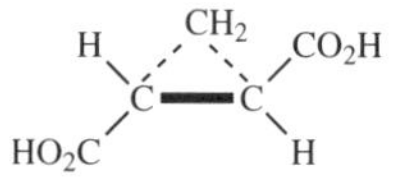

Cyclopropane-1(R), 2(R)-dicarboxylic acid

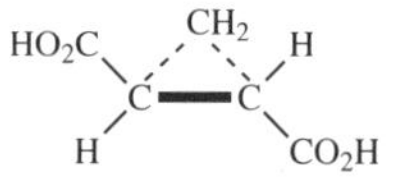

Cyclopropane-1(S), 2(S)-dicarboxylic acid

Enantiomeric *trans*-Cyclopropane-1,2-dicarboxylic acids
melting point = 175°C; pK_1 3.68

for the rotation of polarised light), the *cis* isomer has properties quite different from either of the *trans* isomers. Although the structures of the compounds here are somewhat more complicated, the cyclopropane-1,3-dicarboxylic acids provide another example of the general case of a molecule with two identical chiral centres previously illustrated by the tartaric acids (p. 26). Similar considerations apply to *cis*- and *trans*-1,2- or 1,3-disubstituted cyclohexanes.

Cis–trans isomerism is not confined to compounds with doubly bonded carbon atoms, being found also in compounds with C=N and N=N links.

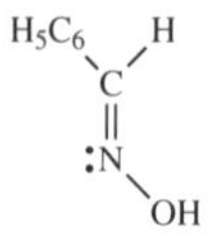

melting point = 34°C 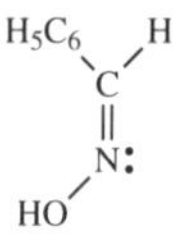   melting point = 127°C   melting point = 68°C 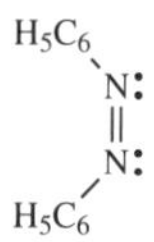   melting point = 71°C

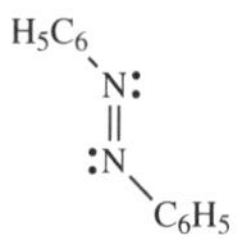

Benzaldoxime   Azobenzene

Examples of isomerism in these compounds are given above, but since the chemistry of these functional groups has not been described in detail, differences between the isomers will not be considered further.

1.4 Mechanism

Armed with all the information covered so far in this chapter we are now almost in a position to consider the course which organic reactions take – the mechanism of the reaction. Before we do this, however, we must first examine some of the factors that affect the distribution of electrons in covalent bonds and their availability for the formation of new bonds.

1.4.1 Electronegativity

Electronegativity is a measure of the ability of an atom of an element to attract the electrons in a bonding pair. Electronegativity generally increases from left to right across the periodic table and decreases on going down any group. The most electronegative elements are therefore found in the top right-hand corner (O, F, Cl) and the least (i.e. most electropositive) in the bottom left-hand corner (Ba, Cs, Rb). A qualitative list of relative electronegativities of some common elements is given below, but it should be realised that the molecular environment may affect electronegativity to a limited extent.

F > O > Cl, N > Br > I, S, C > P, H > B > Mg > Li > Na

1.4.2 The inductive effect

If we consider the distribution of electrons in a single covalent bond between two atoms, it can be seen that the sharing of electrons is not necessarily equal. In a symmetrical molecule A—A (e.g. H_2, Cl_2, HO—OH) the two nuclei, in whose neighbourhood the bonding electrons are found, are indistinguishable, and in the absence of any external effect the electron distribution will be symmetrical (i.e. the electrons will be 'equally shared'). If, however, we consider a molecule A—X where A and X are different (e.g. HF, ICl) then the nuclei are distinct and the atoms A and X may differ greatly in electronegativity. In these circumstances the distribution of electrons may be asymmetric (i.e. 'unequal sharing') and the electron density will be greater near the more electronegative element. This electron displacement carried to an extreme leads to the production of ions, but in many covalent bonds causes only slight **polarisation** of the bond, represented by A→X (not to be confused with A → X, see p. 6) or $A^{\delta+}X^{\delta-}$ where X is the more electronegative and δ+, δ– represent small electrical charges produced by the polarisation. Such a polarised bond may be regarded as a resonance hybrid of the purely covalent bond, in which there is symmetrical electron distribution, and the purely ionic bond in $A^+ X^-$ where both electrons have been transferred entirely to the more electronegative atom. A molecular orbital picture, which is complementary to the resonance description, could indicate the electron density in the bond by shading, in which case the bond in hydrogen fluoride would be represented as shown in (**27**) (Figure 1.20).

Figure 1.20   (**27**)

In a slightly more complex group of atoms such as A—A—X, the electron distribution in the bond A—A will not be completely symmetrical, as the polarisation of the A—X bond produces a partial positive charge on the central atom, i.e. A—A$^{\delta+}$X$^{\delta-}$. This charge makes the central atom slightly more electronegative than its left-hand neighbour, resulting in another smaller electron displacement, i.e. A$^{\delta\delta+}$ $^{\delta\delta-}$A$^{\delta+}$ $^{\delta-}$X which might be represented by A→A→→X. This transmission of polarisation to adjacent bonds is known as the inductive effect, and because the electrons of σ bonds are localised (p. 6), the inductive effect in a chain of singly bonded atoms dies away very rapidly indeed.

1.4.3 The mesomeric or conjugative effect

The effect of relative electronegativity on the electron distribution in a double bond is very similar to that described above for the case of a single bond, except that the diffuse electron clouds of the π bond can be polarised to a very much greater extent. In the carbonyl group $\diagdown$C=O this can be regarded as the molecular orbital with asymmetric electron distribution as in Figure 1.21 (**28**). The polarisation of the σ bond in these circumstances is insignificant and can be ignored.

Figure 1.21   (**28**)

The polarisation of π bonds differs from that in σ bonds not only by its magnitude, but also by the extent to which it can be transmitted. The interaction of adjacent double bonds in a conjugated system has already been described (p. 11) and if such a conjugated system contains an electronegative atom then very extensive electron displacement may occur along the conjugated chain of double bonds. In the case of propenal (CH_2=CH—CH=O) the molecular structure can be conventionally represented as (**29**), but on account of the electronegativity of the oxygen atom the correct description of the molecule is a resonance hybrid of the three canonical structures (**29a, b, c**). The contribution from (**29c**) must be considered, as the partial positive charge on carbon atom 1 will polarise the electron clouds of the

$$CH_2=CH-CH=\ddot{\underset{..}{O}}: \longleftrightarrow CH_2=CH-\overset{+}{CH}-\ddot{\underset{..}{O}}:^- \longleftrightarrow \overset{+}{CH_2}-CH=CH-\ddot{\underset{..}{O}}:^-$$

(29a)   (29b)   (29c)

adjacent π bond very considerably. The molecular orbital picture of this system is simply an asymmetric distribution of electrons in the extended molecular orbitals of the conjugated system, one of which is illustrated in Figure 1.22 (**30**). The transmission of electron displacement in conjugated double bond systems is known as the **mesomeric** (or **conjugative**) **effect**, and differs from the inductive effect both in the greater polarisation possible with π bonds, and the much greater distance over which the polarisation can be transmitted.

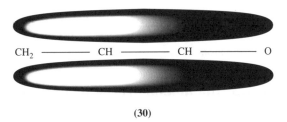

$CH_2 \longrightarrow CH \longrightarrow CH \longrightarrow O$

Figure 1.22   (30)

1.4.4 The breaking and formation of covalent bonds

Almost all reactions in organic chemistry involve the breaking or formation of covalent bonds, and an appreciation of how these processes can occur is fundamental to the understanding of reaction mechanisms. If two atoms are joined by a covalent bond, e.g. A—B, and during a reaction this bond is broken, then there are three ways in which this can occur, distinguished by the fate of the bonding electrons. **Homolytic fission** is the name given to the breaking of a covalent bond, in which each atom retains one of the shared electrons. Each of the species produced has an odd or 'unpaired' electron, but bears no electrical charge, and is known as a **radical**. **Heterolytic fission** is the breaking of a covalent bond in which both of the bonding electrons remain with one of the atoms. This process, which produces electrically charged species, can occur in two ways, as illustrated below. Clearly if B is significantly more electronegative than A, the latter mode of heterolytic fission is more probable.

$$A\frown\!\!\frown B \longrightarrow A\cdot + \cdot B \qquad \text{Homolytic fission}$$
$$A\frown B \longrightarrow A:^- + B^+$$
$$A\frown B \longrightarrow A^+ + :B^- \quad \bigg\}\text{Heterolytic fission}$$

The formation of covalent bonds can occur by exactly the reverse of the processes described above, and although organic chemistry contains reactions that involve more complex processes, a great many reactions can be explained by the use of these simple electron movements. It is convenient, when attempting to represent a reaction mechanism graphically, to use symbols indicating the direction of movement of electrons. The only two such symbols employed in this book are the curly arrows ⌢ and ⌢ which indicate the movement of one and two electrons respectively. When using these symbols, students

$$A\!-\!B \longrightarrow A\cdot + B\cdot$$
$$A\!-\!B \longrightarrow A^{\oplus} + :B^{\ominus}$$

should ensure that the source and destination of the electrons are clearly indicated by the positions of the root and tip of the arrow, e.g. the reaction of ammonia and hydrogen chloride can be represented:

$$H_3N: \curvearrowright H\!-\!\ddot{C}\ddot{l}: \longrightarrow H_3\overset{\oplus}{N}\!-\!H + :\ddot{\underset{..}{C}}\ddot{l}:^{\ominus}$$

the lone-pair of the nitrogen being donated to form a bond to the hydrogen atom, with simultaneous transfer of the electrons of the H—Cl bond to the chlorine atom.

1.4.5 The mechanism of complex reactions

We now consider the mechanism of some common reactions in detail, and it is convenient to classify the reactions into three groups, with obvious relationships to the method of bond formation and breakage described earlier. We will use examples to illustrate important mechanistic features, e.g. formation and stability of intermediates, and regioselectivity.

Radical reactions
As mentioned previously, radicals are species with an odd number of electrons. They are normally very reactive, and attack molecules at positions of high electron density. The atoms of hydrogen, bromine and chlorine are simple examples of radicals. Although of relatively little importance in terms of the chemistry of biological systems, we will study them first as you may already be familiar with the mechanism of a radical reaction, the halogenation of alkanes.

Halogenation of alkanes. Methane and chlorine undergo a very rapid reaction, at high temperatures (> 250 °C) or under the influence of light, and this is an example of a radical process. In this process several thousand product molecules are produced for each photon of light absorbed. This is typical of a chain reaction and the mechanism that best accounts for these facts comprises three steps.

1. **Initiation.** Chlorine molecules absorb light and dissociate into atoms:

$$:\!\ddot{\underset{..}{C}}\ddot{l}\!-\!\ddot{\underset{..}{C}}\ddot{l}: \xrightarrow{h\nu} 2\,:\!\ddot{\underset{..}{C}}\ddot{l}\cdot$$

Photolysis of a weak bond is a common means of producing radicals. Diacyl peroxides are a common source of carbon-based radicals, which can be obtained by either photolysis or thermolysis;

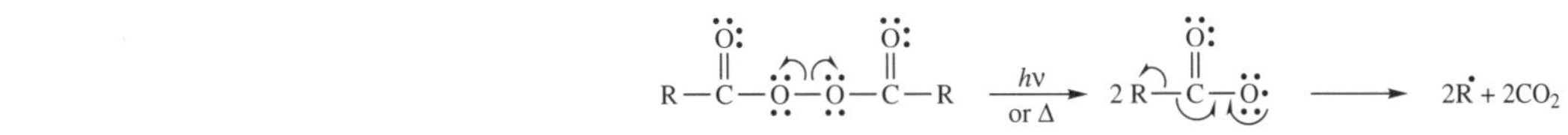

$$\text{R}-\overset{\overset{\ddot{\text{O}}:}{\|}}{\text{C}}-\overset{..}{\underset{..}{\text{O}}}-\overset{..}{\underset{..}{\text{O}}}-\overset{\overset{\ddot{\text{O}}:}{\|}}{\text{C}}-\text{R} \quad \xrightarrow[\text{or } \Delta]{h\nu} \quad 2\,\text{R}-\overset{\overset{\ddot{\text{O}}:}{\|}}{\text{C}}-\overset{..}{\underset{..}{\text{O}}}\cdot \quad \longrightarrow \quad 2\text{R}\cdot + 2\text{CO}_2$$

2. **Propagation.** A chlorine atom can now abstract a hydrogen atom from an alkane molecule to form hydrogen chloride and an alkyl (methyl) radical:

$$:\!\overset{..}{\underset{..}{\text{Cl}}}\!\cdot \quad \text{H}-\text{CH}_3 \quad \longrightarrow \quad \text{H}-\overset{..}{\underset{..}{\text{Cl}}}: \;+\; \cdot\text{CH}_3$$

The alkyl radical can then attack a chlorine molecule producing a molecule of methyl chloride and a chlorine atom, which can continue the chain reaction by reacting with another alkane molecule:

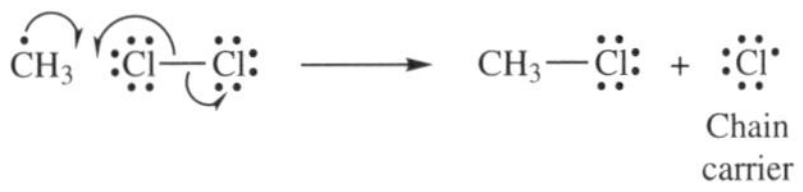

$$\cdot\text{CH}_3 \quad :\!\overset{..}{\underset{..}{\text{Cl}}}\!-\!\overset{..}{\underset{..}{\text{Cl}}}: \quad \longrightarrow \quad \text{CH}_3-\overset{..}{\underset{..}{\text{Cl}}}: \;+\; :\!\overset{..}{\underset{..}{\text{Cl}}}\!\cdot$$
Chain carrier

3. **Termination.** The chain reaction can be terminated by the combination of any two of the radicals involved:

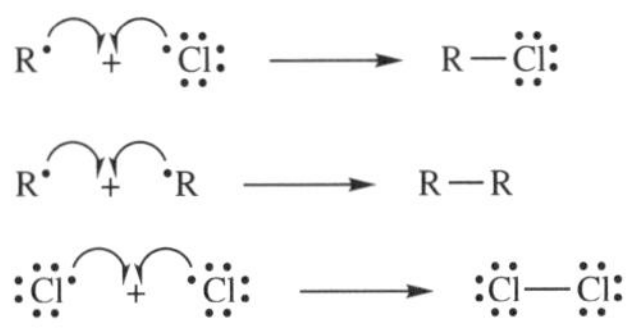

$$\text{R}\cdot \;+\; \cdot\overset{..}{\underset{..}{\text{Cl}}}: \quad \longrightarrow \quad \text{R}-\overset{..}{\underset{..}{\text{Cl}}}:$$

$$\text{R}\cdot \;+\; \cdot\text{R} \quad \longrightarrow \quad \text{R}-\text{R}$$

$$:\!\overset{..}{\underset{..}{\text{Cl}}}\!\cdot \;+\; \cdot\overset{..}{\underset{..}{\text{Cl}}}: \quad \longrightarrow \quad :\!\overset{..}{\underset{..}{\text{Cl}}}\!-\!\overset{..}{\underset{..}{\text{Cl}}}:$$

(Note that although the first termination reaction shown above leads to product, it removes radicals from the chain reaction. If all product was formed from the combination of two radicals, each photon of light would give only one product molecule!) It is not generally possible to obtain solely monosubstituted products from this reaction, as the alkyl halides are further attacked by chlorine atoms.

Similar reactions occur with bromine and alkanes. Fluorine reacts extremely violently but iodination is possible only in the presence of an oxidising agent (to remove the HI formed). The halogenation of alkanes is *not* suitable for the laboratory preparation of alkyl chlorides owing to the mixtures obtained. In addition, the bromination/chlorination of the higher alkanes will give all of the possible isomeric products (since the relative stability of radicals is not greatly different) but those products formed via the most stable radical will predominate, e.g.

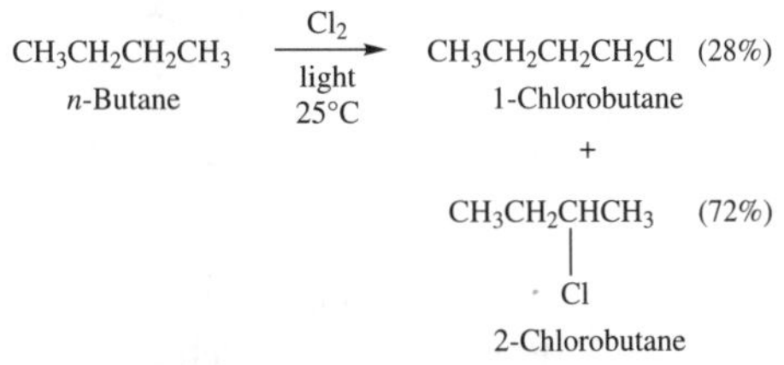

$$\underset{n\text{-Butane}}{\text{CH}_3\text{CH}_2\text{CH}_2\text{CH}_3} \quad \xrightarrow[\substack{\text{light}\\25°\text{C}}]{\text{Cl}_2} \quad \underset{\text{1-Chlorobutane}}{\text{CH}_3\text{CH}_2\text{CH}_2\text{CH}_2\text{Cl}} \; (28\%)$$

$$+$$

$$\underset{\text{2-Chlorobutane}}{\text{CH}_3\text{CH}_2\overset{\displaystyle |}{\underset{\displaystyle \text{Cl}}{\text{CH}}}\text{CH}_3} \; (72\%)$$

Simple radicals, e.g. $R_3C\cdot$, are planar with the central carbon atom sp^2 hybridised and the unpaired electron in the unhybridised p orbital:

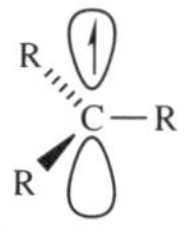

The relative stability of alkyl free radicals reflects the relative ease with which the C—H bond will undergo homolytic fission:

$$R_3\dot{C} \quad > \quad R_2\dot{C}H \quad > \quad R\dot{C}H_2 \quad > \quad \dot{C}H_3$$

Tertiary (3°)   Secondary (2°)   Primary (1°)

Most stable   Least stable
(most easily formed)   (least easily formed)

$$R_3C-H \quad < \quad R_2CH-H \quad < \quad RCH_2-H \quad < \quad CH_3-H$$

Weakest CH bond   Strongest

This order of stability also agrees with decreasing relief of strain (if R is large) on going from a sterically crowded sp³ hybridised alkane to a less crowded sp² hybridised radical.

Benzyl and allyl radicals are more stable than simple alkyl radicals owing to delocalisation:

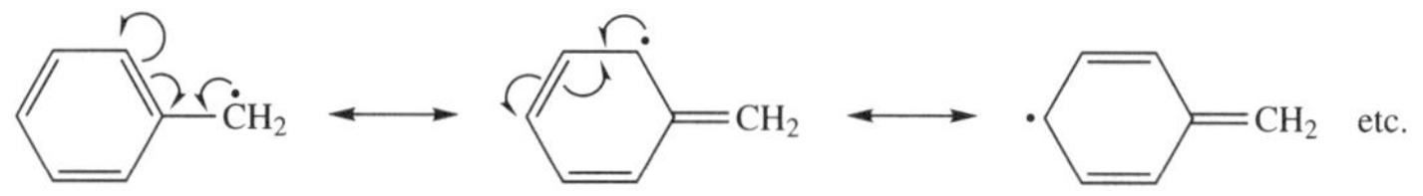

Benzyl radical

Resonance hybrid

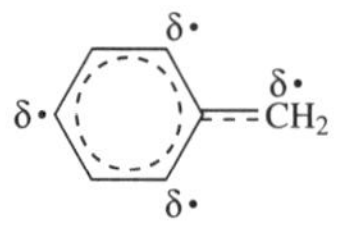

Allyl radical

Resonance hybrid

Hydrobromination of alkenes in sunlight. In sunlight, or in the presence of peroxides, the addition of HBr to alkenes proceeds via a rapid chain reaction and the most stable free radical:

1. **Initiation.**

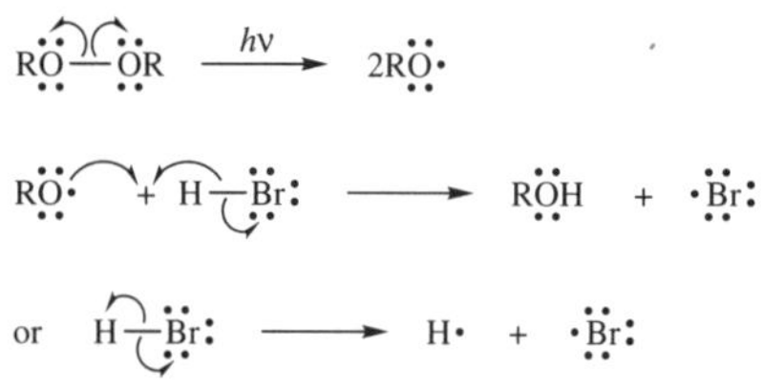

$$RO-OR \xrightarrow{h\nu} 2RO\cdot$$

$$RO\cdot + H-Br: \longrightarrow ROH + \cdot Br:$$

or $\quad H-Br: \longrightarrow H\cdot + \cdot Br:$

2. **Propagation.** The bromine atom, the chain carrier, now adds to the alkene to give the most stable radical:

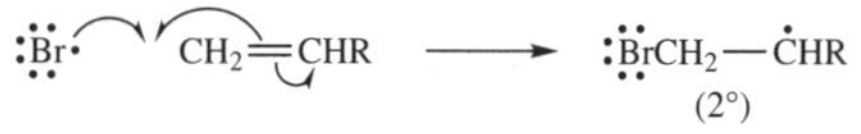

$$:\!Br\cdot + CH_2=CHR \longrightarrow :\!BrCH_2-\dot{C}HR$$
$$(2°)$$

not

$$:\ddot{Br}\cdot \quad RCH{=}CH_2 \longrightarrow :\ddot{Br}CH{-}\overset{R}{\underset{|}{CH_2}}$$
(1°)

The alkyl radical can now abstract a proton from HBr to give the product and a bromine atom:

$$BrCH_2{-}\dot{C}HR + H{-}Br \longrightarrow BrCH_2{-}CH_2R + \cdot\ddot{Br}:$$

e.g. $CH_3CH{=}CH_2 + HBr \xrightarrow{\text{Peroxides}} CH_3CH_2CH_2Br$

Propene   1-Bromopropane
(major product)

The product obtained is known as the anti-Markownikoff product.

3. **Termination.** Once again, chain termination involves the combination of any two radicals, e.g.

$$R\dot{C}H{-}CH_2Br + \cdot\ddot{Br}: \longrightarrow R\underset{\underset{Br}{|}}{CH}{-}CH_2Br$$

$$:\ddot{Br}\cdot + \cdot\ddot{Br}: \longrightarrow :\ddot{Br}{-}\ddot{Br}:$$

Electrophilic reactions

Electrophiles are species with a deficiency of electrons, e.g. a vacant atomic orbital, sometimes bearing a positive charge. They are attacked by positions of high electron density or negative charge. Examples of electrophiles are H+, $AlCl_3$, R_3C^+.

One of the most important electrophilic reactions is the addition to double bonds, e.g. C=C or C=O. For example, the first step in the acid-catalysed addition of nucleophiles to carbonyl groups (see later) involves the electrophilic addition of H+:

$$\overset{\delta+}{C}{=}\overset{\delta-}{\ddot{O}} \xrightarrow{H^{\oplus}} \rightleftharpoons \overset{}{C}{=}\overset{\oplus}{O}H \longleftrightarrow \overset{\oplus}{C}{-}\ddot{O}H$$

(note the double-headed arrow indicating the movement of two electrons). The effect of protonation on the oxygen of the carbonyl group is to increase the electrophilicity of the carbonyl group, i.e. increase the positive charge on the carbon atom, making it more reactive towards weaker nucleophiles.

Addition to alkenes. Simple carbocations must be planar (sp² hybridised) and have a vacant, unhybridised p orbital:

The simple alkyl carbocations follow the stability sequence;

$$R_3C^{\oplus} > R_2\overset{\oplus}{C}H > R\overset{\oplus}{C}H_2 > \overset{\oplus}{C}H_3$$
$$3° \quad\quad\quad 2° \quad\quad\quad 1°$$

Most stable — easiest to form   Least stable — most difficult to form

Tertiary carbocations, R_3C^+, are stabilised by the electron donation from three alkyl groups, which effectively reduces the positive charge and stabilises the cation. This effect is obviously greater for tertiary carbocations than for secondary (only two alkyl groups):

$$R \rightarrow \overset{\oplus}{C}(R)(R)$$

The order of carbocation stability is the same as that of radical stability ($3° > 2° > 1° > CH_3$) and, once again, delocalisation (the spreading out) of the positive charge helps produce more stable carbocations, e.g. benzyl, allyl. A further possibility is carbocation stabilisation by a lone-pair on an adjacent atom, e.g.

$$CH_3\ddot{\text{O}}\frown\overset{\oplus}{C}H_2 \longleftrightarrow CH_3\overset{\oplus}{\ddot{\text{O}}}=CH_2$$

Electrophilic addition to an alkene can be represented by:

$$\text{C}=\text{C} + \overset{\delta+}{\text{X}}-\overset{\delta-}{\text{Y}} \longrightarrow \underset{\text{X}}{\text{C}}-\underset{\text{Y}}{\text{C}}$$

Here Y is more electronegative than X (so that XY is polarised) and initially electrons are donated from the electron-rich π bond to the electrophilic (electron-deficient) X atom (or group) to form an intermediate carbocation and an anion Y⁻:

$$\underset{H}{\overset{H}{\text{C}}}=\underset{H}{\overset{H}{\text{C}}} \quad\quad \longrightarrow \quad \overset{\oplus}{C}H_2-CH_2-X \quad \longrightarrow \quad CH_2(Y)-CH_2(X)$$
$$\overset{\delta+}{X}-\overset{\delta-}{Y} \quad\quad\quad\quad\quad \ddot{Y}^{\ominus}$$

The carbocation reacts immediately with the anion to form the addition compound. The driving force for this reaction is the replacement of a weak π bond by two strong σ bonds (to X and Y), making this an energetically favourable process.

Addition reactions of this type are:

$$\underset{H}{\overset{H}{>}}C=C\underset{H}{\overset{H}{<}} \;+\; \begin{cases} \overset{\delta+\;\;\delta-}{X-Y} \\ H-Cl \longrightarrow H-\underset{\underset{H}{|}}{\overset{\overset{H}{|}}{C}}-\underset{\underset{Cl}{|}}{\overset{\overset{H}{|}}{C}}-H \\ H-Br \longrightarrow H-\underset{\underset{H}{|}}{\overset{\overset{H}{|}}{C}}-\underset{\underset{Br}{|}}{\overset{\overset{H}{|}}{C}}-H \\ H-I \longrightarrow H-\underset{\underset{H}{|}}{\overset{\overset{H}{|}}{C}}-\underset{\underset{I}{|}}{\overset{\overset{H}{|}}{C}}-H \\ \underset{(Cl_2+H_2O)}{Cl-OH} \longrightarrow H-\underset{\underset{Cl}{|}}{\overset{\overset{H}{|}}{C}}-\underset{\underset{OH}{|}}{\overset{\overset{H}{|}}{C}}-H \;\;\text{(a 'chlorohydrin')} \\ \underset{(Br_2+H_2O)}{Br-OH} \longrightarrow H-\underset{\underset{Br}{|}}{\overset{\overset{H}{|}}{C}}-\underset{\underset{OH}{|}}{\overset{\overset{H}{|}}{C}}-H \;\;\text{(a 'bromohydrin')} \\ H-OSO_3H \longrightarrow H-\underset{\underset{H}{|}}{\overset{\overset{H}{|}}{C}}-\underset{\underset{OSO_3H}{|}}{\overset{\overset{H}{|}}{C}}-H \xrightarrow{+H_2O} H-\underset{\underset{H}{|}}{\overset{\overset{H}{|}}{C}}-\underset{\underset{OH}{|}}{\overset{\overset{H}{|}}{C}}-H + H_2SO_4 \\ Cl-Cl \longrightarrow H-\underset{\underset{Cl}{|}}{\overset{\overset{H}{|}}{C}}-\underset{\underset{Cl}{|}}{\overset{\overset{H}{|}}{C}}-H \;\;\text{(and addition of } Br_2\text{)} \end{cases}$$

In these reactions, if an unsymmetrical molecule adds across an unsymmetrical alkene, two products are possible but one is usually formed in preference to the other. The two possible reaction sequences are:

$$\underset{H-X}{\overset{R}{>}}C=C\underset{H}{\overset{H}{<}} \longrightarrow \begin{cases} R\overset{\oplus}{CH}-CH_3 \;\; \xrightarrow{\overset{\ominus}{X}} \;\; R\underset{|}{\overset{X}{C}}HCH_3 \;\text{(major)} \\ (2°) \\ R\overset{\oplus}{CH_2CH_2} \;\; \xrightarrow{\overset{\ominus}{X}} \;\; RCH_2CH_2X \\ (1°) \end{cases}$$

The factor that determines the course of this reaction is the ease of formation of the alternative carbocation intermediates. Thus, the product formed in an electrophilic addition to an alkene is the one formed via the most stable carbocation, i.e. R—CHX—CH$_3$ above. This is summarised by the Markownikoff rule: 'In an addition reaction of an unsymmetrical alkene the more positive end (atom or group) adds to the carbon atom bearing the greater number of hydrogen atoms' (often quoted as 'them that has gets'), e.g.

$$CH_3-CH=CH_2 \;+\; \overset{\delta-}{H}O-\overset{\delta+}{C}l \longrightarrow CH_3\underset{\underset{OH}{|}}{C}H-\underset{\underset{Cl}{|}}{C}H_2$$

$$(\text{via } CH_3\overset{\oplus}{C}H-CH_2Cl)$$

Mechanism and stereochemistry of addition to alkenes

Attention here will be directed to the addition of hydrogen halides, HX, to alkenes, although the same principles apply in many other cases.

Addition reactions initiated by nucleophilic attack on alkenes (i) are restricted to cases where the alkene is conjugated to an electronegative group, i.e. the Michael addition. In these cases the intermediate carbanion is usually a fairly stable species (e.g. an enolate anion) so that rotation can occur about the single bonds joined to the anionic centre and the subsequent attachment of a proton does not occur with any special orientation with respect to the new C—X bond.

The second process is probably the best simple description of the mechanism of addition to alkenes, although the timing of the two steps may vary. If a very stable carbocation (i.e. tertiary) is formed in the first step, this may have an appreciable lifetime before reaction with the nucleophile. With reactions in which secondary or primary carbocations would be formed, the intermediate carbocation scarcely has any independent existence as the two steps follow in rapid succession, approaching the timing of a concerted addition process (iii). In these cases the lifetime of the carbocation is so short that bond rotation is not possible, and it is found that H$^+$ and X$^-$ add to

the double bond in an *anti* orientation. *Syn*-additions to alkenes are known in two very similar cases. Alkenes are converted into *vic*-diols by reaction with alkaline solutions of permanganates or osmium tetroxide. Both MnO_4^- and OsO_4 form cyclic intermediates whose decomposition results in the formation of diols, the overall reaction being a *syn*-addition of two hydroxyl groups.

(*Syn*- and *anti*- have meanings similar to *cis*- and *trans*- but are used to describe stereochemical relationships between reagents during reaction, whereas the latter are reserved for description of molecular structure.)

Many alkenes can exist as *cis–trans* isomers (p. 33) and the stereochemical requirements of addition to alkenes (when long-lived carbocations are not intermediates) lead to important results, which can be illustrated by the addition of bromine to *cis–trans* isomeric alkenes. Halogen addition is an *anti*-addition, similar to those described above, and *anti*-addition of bromine can occur in two ways to Z-but-2-ene and in two ways to E-but-2-ene. The products from the addition to the Z-alkene are enantiomers, but the alternative additions to the E-alkene give identical *meso* products. Analogous results arise from stereospecific *syn*-addition (e.g. hydroxylation by osmium tetroxide).

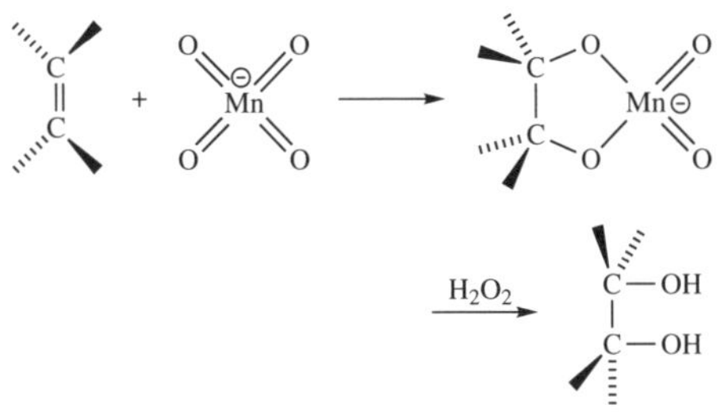

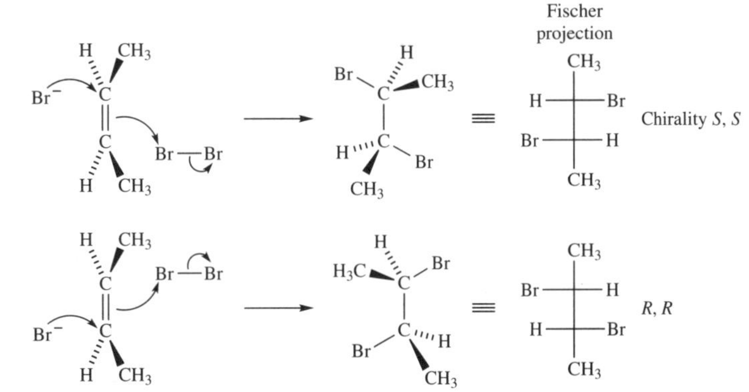

Anti-additions to Z-but-2-ene

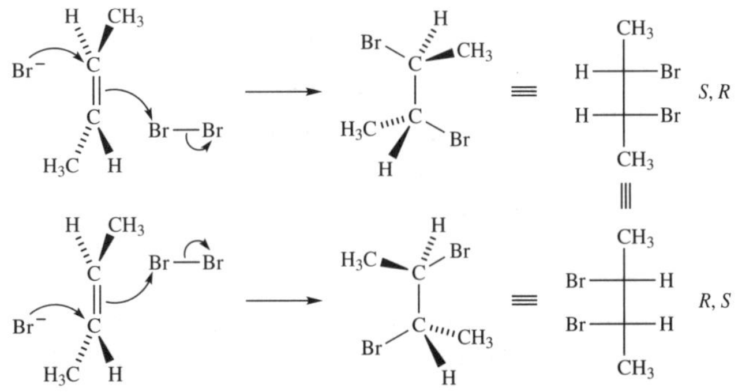

Anti-additions to E-but-2-ene

Note that if an unsymmetrical alkene (e.g. pent-2-ene) were to be used, then *anti*-addition of halogen to the Z-isomer would give one pair of enantiomeric products and *anti*-addition to the E-isomer would give another pair of enantiomers.

Electrophilic aromatic substitution

Electrophilic aromatic substitution is of little biological significance but is a reaction with which you are probably already familiar, e.g. nitration of benzene by the nitronium ion electrophile, NO_2^+:

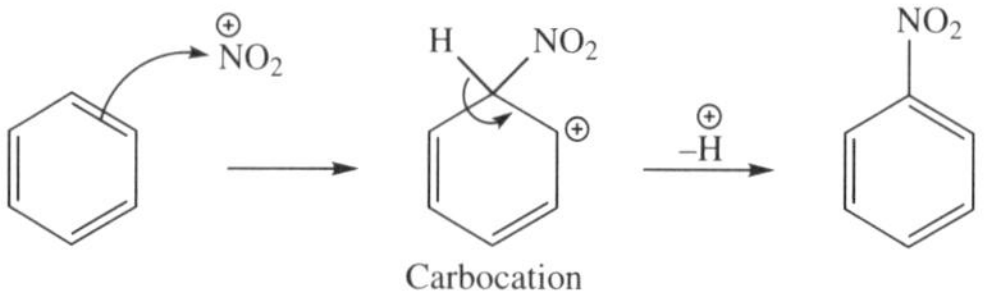

Nucleophilic reactions

Nucleophiles are species with a pair of electrons (e.g. a lone-pair) available for bond formation, sometimes bearing a negative charge, which attack positions of low electron density or positive charge. Examples of nucleophiles are OH^-, I^-, H_2O and NH_3. Important examples of the reactions of nucleophiles are nucleophilic attack at saturated and unsaturated carbon atoms (covered in detail in Chapter 3).

Nucleophilic attack at a saturated carbon atom

Detailed kinetic studies, i.e. the effects of concentration, temperature and solvent, of reactions of the type

$$R_3C-X + Y^\ominus \longrightarrow R_3C-Y + X^\ominus$$

$$\text{e.g. } CH_3-I + \overset{\ominus}{C}N \longrightarrow CH_3CN + I^\ominus$$

show that the rate equation takes one of two forms:

1. Rate = $k_2[R_3CX][Y^-]$, known as **S_N2**, or
2. Rate = $k_1[R_3CX]$, known as **S_N1**.

Nucleophilic substitution bimolecular, S_N2

$$\text{Rate} = k_2[R_3CX][Y^-]$$

The rate data for S_N2 reactions suggest that both the nucleophile and the alkyl halide participate in the rate-determining process, and this is interpreted as meaning that the departure of the leaving group, X^-, is assisted by the approach of the nucleophile, the reaction occurring via a transition state (activated complex) in which the R_3C-X bond is partly broken and the R_3C-Y bond partly formed.

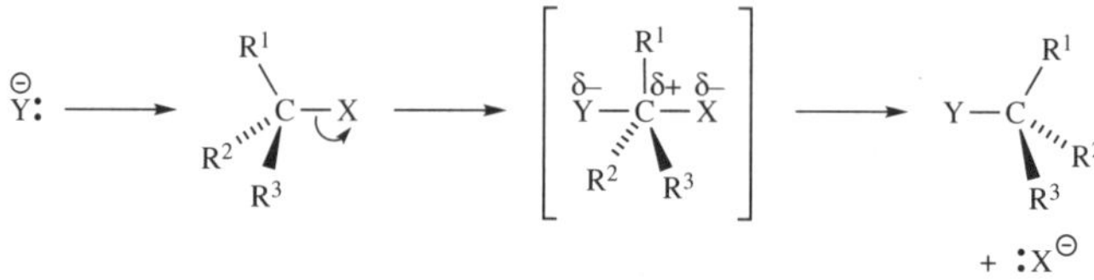

In the transition state, the central carbon atom is sp^2 hybridised (planar), with the nucleophile, Y, and leaving group, X, associated with the lobes of the unhybridised p orbital. The original negative charge of the nucleophile, Y⁻, is now shared between the nucleophile and the leaving group, both of which have a partial negative charge. The central carbon atom carries a partial positive charge.

The transition state is more sterically crowded than the substrate (surrounded by five rather than four groups). This increase in crowding will increase as the size of the original substituents increases, so we would expect the S_N2 mechanism to be more favoured for less substituted substrates (CH_3X, 1°) and less favoured for more substituted substrates (3°). The energy profile for an S_N2 reaction consists of a single energy barrier, with the transition state representing the point of highest energy (Figure 1.23).

All nucleophilic substitutions following S_N2 kinetics proceed with inversion of the configuration of the central carbon atom, known as **Walden inversion**. This inversion is a direct result of the nucleophile attacking the face of the substrate remote from the leaving group – this minimises the repulsion between the similarly charged nucleophile and leaving group. If the substrate in an S_N2 reaction is chiral ($R^1R^2R^3CX$) then the product will also be chiral (no loss of optical activity – remember, however, that there is no simple relationship between absolute configuration and direction of rotation of light).

Nucleophilic substitution unimolecular, S_N1

Rate = k_1 [R_3CX]

The rate equation for the S_N1 reaction is taken to imply that the rate-determining step (RDS), which does not involve the nucleophile, is the slow ionisation of the C—X bond, which is followed by the very much more rapid reaction of the resultant carbocation with the nucleophile:

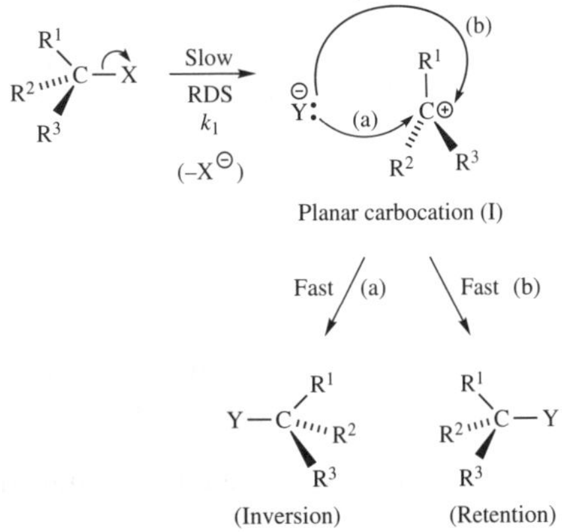

Attack of the nucleophile, Y⁻, on the planar (sp² hybridised) carbocation can occur, with equal probability, at either face to give (a) the product of retention of configuration, or (b) the product of inversion of configuration. These products, of (a) and (b), are enantiomers so that, if the substrate ($R^1R^2R^3CX$) was chiral, the result of an S_N1 reaction would be racemisation. (This is the ideal case; in reality the leaving group, X⁻, often remains associated with the

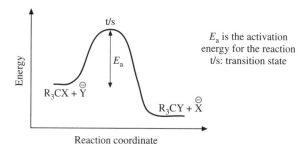

Figure 1.23

face of the substrate from which it is leaving, thus hindering nucleophilic attack on this face.)

The energy profile for an S_N1 reaction consists of two energy barriers (each with an associated transition state) and a reaction intermediate, the planar carbocation (Figure 1.24). The S_N1 mechanism, then, is characterised by a carbocation intermediate so the more stable this intermediate, the more likely this mechanism is to operate. As mentioned previously, the order of carbocation stability is:

3°, allyl, benzyl > 2° > 1° > CH_3

hence, 3°, allyl and benzyl substrates undergo nucleophilic substitution via an S_N1 mechanism.

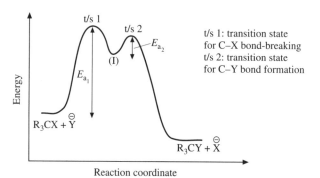

Figure 1.24

1.4.6 The mechanism and stereochemistry of alkene-forming eliminations

The general pattern of elimination reactions may be represented by:

$$-\underset{X}{\underset{|}{C}}-\underset{Y}{\underset{|}{C}}- \xrightarrow{-XY} \diagdown_{C=C}\diagup$$

We will confine our attention to eliminations in which Y is hydrogen (although the same stereochemical principles apply in many other cases). In principle, the elimination of HX could occur by three pathways:

In the first two processes, elimination occurs in a stepwise fashion, either H^+ or X^- being lost initially. The first type of elimination, via carbanion formation, is comparatively rare and will not be considered further. In general, this type of reaction requires that the initial carbanion be stabilised in some way, e.g. by conjugation with an adjacent carbonyl group. The second type of elimination starts by initial formation of a carbocation followed by loss of a proton. This is likely to occur only if the intermediate carbocation is tertiary and it should be noted that this ionisation is identical with the first stage of an S_N1 reaction. It is observed, in practice, that alkene formation is an important side reaction during nucleophilic substitution of tertiary alkyl halides, though insignificant in S_N2 reactions. The third pathway for elimination involves a concerted attack of base on the proton and loss of X^-. This is the most important elimination pathway for primary and secondary alkyl halides. (Several examples of eliminations will be found on p. 52.)

If the ionic intermediates in schemes (i) and (ii) have lifetimes long enough to permit rotation about single bonds before the second stage of elimination ensues, then no special steric requirement will be observed for the relative orientation of the C—H and C—X bonds in the starting material. However, it is found that in the concerted (E2) elimination (iii), the reaction proceeds most readily if the C—H and C—X bonds are 'antiperiplanar', i.e. lie in the same plane and on opposite sides of the common C—C bond. Elimination is not impossible if the relevant bonds are otherwise oriented, but the rate of elimination may be decreased by a factor of several thousand.

Antiperiplanar orientation of C–H and C–X bonds

The evidence for the antiperiplanar orientation required for a concerted elimination is supported by experiments with rigid systems such as steroids in which the bonds are held in fixed relative positions. Similar results are observed in simpler systems. Thus menthyl chloride reacts with sodium ethoxide to give one substituted cyclohexene exclusively, since in neither of the chair conformations is it possible to get the C—H and C—Cl bonds in the antiperiplanar arrangement required for elimination in the other direction.

Note that in cyclic systems such as those shown above, the elimination requires not only that the leaving groups are formally *trans* about a common C—C bond in a planar diagram, but, to achieve the necessary coplanarity both bonds must be axial in the conformation from which elimination occurs.

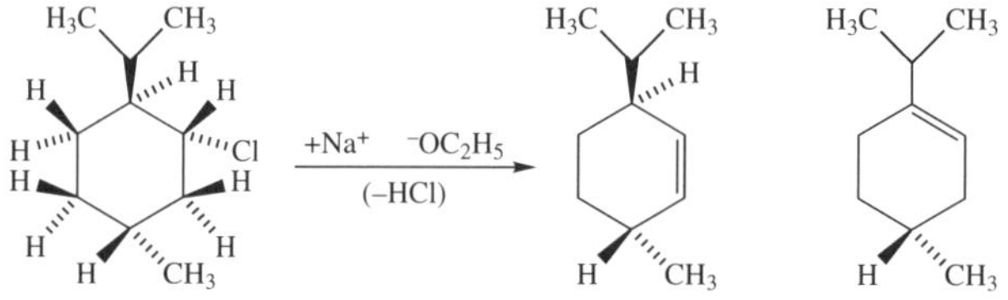

Menthyl chloride

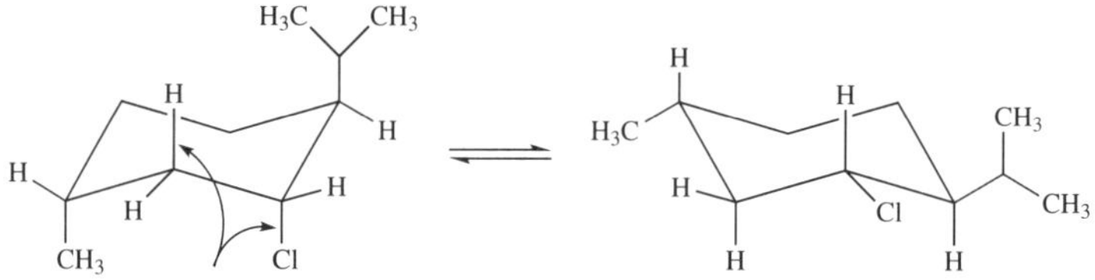

Axial C–H and C–Cl bonds in antiperiplanar orientation

The equatorial C–Cl bond has no antiperiplanar C–H bond in this conformation

1.5 Summary

1. **Hybridisation.** Carbon bonded to four other atoms/groups is sp³ hybridised and the four sp³ hybrid orbitals are arranged about the nucleus with tetrahedral symmetry (109°). The carbon of a double bond is sp² hybridised and the three sp² orbitals are arranged symmetrically in a plane, with an angle of 120° between them. The remaining p orbital can form a π bond with a parallel p orbital on an adjacent carbon atom. Finally, the carbon of a triple bond is sp hybridised with the two sp orbitals arranged linearly (180°).

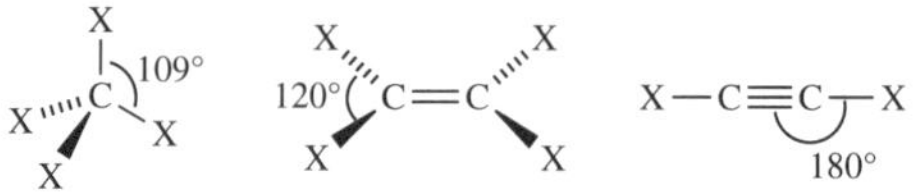

2. **Resonance.** The greater the number of resonance structures which can be drawn for a structure, the more stable that structure is. The true structure will be the average of the resonance (canonical) forms and is called the resonance hybrid.

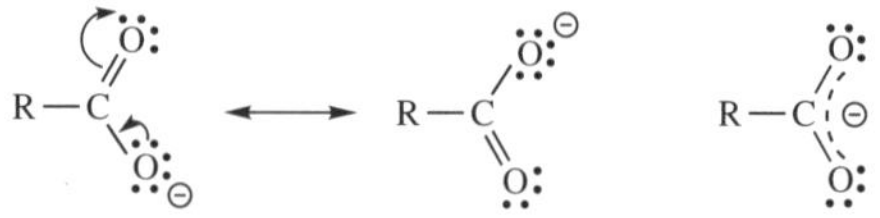

Carboxylate anion   Hybrid

3. **Stereochemistry – enantiomerism.** A structure containing a carbon attached to four different groups is non-superimposable on its mirror image and is said to be chiral. A compound with one chiral centre has two non-superimposable mirror-image isomers known as enantiomers.

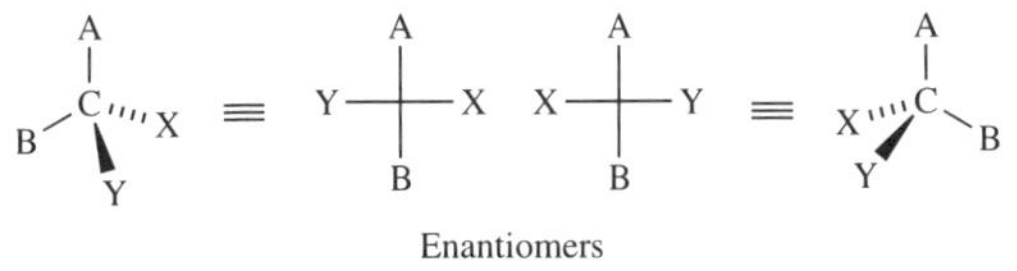

Enantiomers

Stereoisomers that are not mirror images are known as diastereoisomers. Unlike enantiomers, which differ only in their optical rotation and reactions with other chiral molecules, diastereoisomers have different physical properties. Molecules with two, or more, chiral centres can be enantiomers or diastereoisomers.

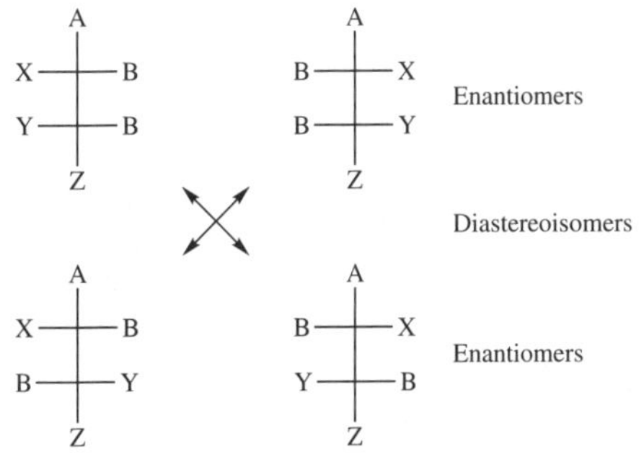

Enantiomers

Diastereoisomers

Enantiomers

The two-dimensional representation of molecules with chiral centres is usually done by Fischer projections and chirality is designated by the Cahn–Ingold–Prelog rules.

4. **Stereochemistry – conformation and geometrical isomerism.** Conformations can be interconverted by rotation (about single bonds). Configurations cannot be interconverted by rotation.

Conformations of *n*-butane

trans (E)   cis (Z)

Configurations of but-2-ene

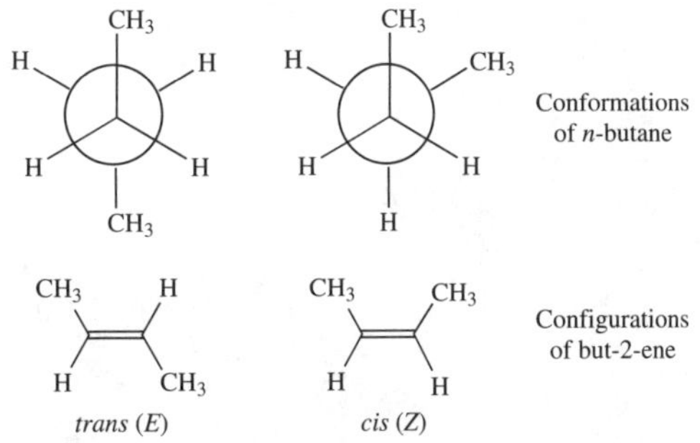

5. **Mechanism.** The movement of electrons in mechanistic schemes is represented by arrows. A single-headed arrow represents the movement of one electron (i.e. a radical reaction). A double-headed arrow represents the movement of two electrons. Always remember that the electrons are

moving so the arrow should start on the site of high electron density (nucleophile) and finish on the site of low electron density (electrophile).

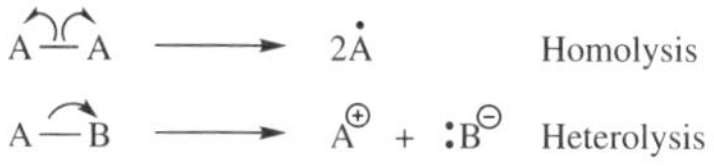

$$A\frown A \longrightarrow 2A\cdot \qquad \text{Homolysis}$$

$$A\frown B \longrightarrow A^{\oplus} + :B^{\ominus} \qquad \text{Heterolysis}$$

Problems

1.1 The energy level diagram for a first row diatomic molecule is shown below. Label the orbitals in the individual atoms and, by considering how these orbitals overlap, label the molecular orbitals. Use this diagram to describe the bonding in dinitrogen (N_2), difluorine (F_2) and dilithium (Li_2) – as in *Star Trek*.

X X$_2$ X

1.2 Give the hybridisation of all the asterisked carbon atoms in Enovid (the first oral contraceptive):

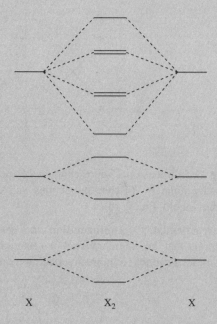

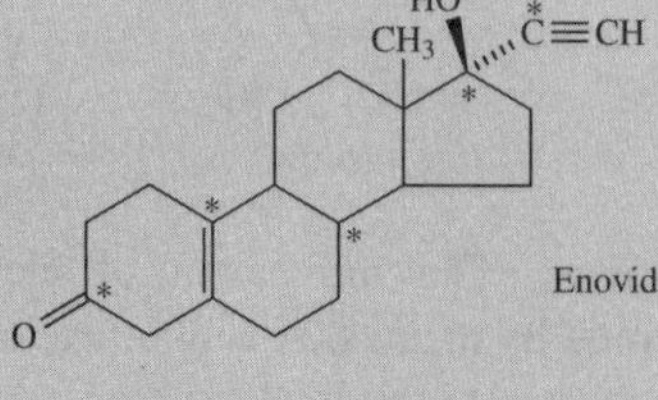

Enovid

1.3 Draw all the possible resonance forms for the following species:

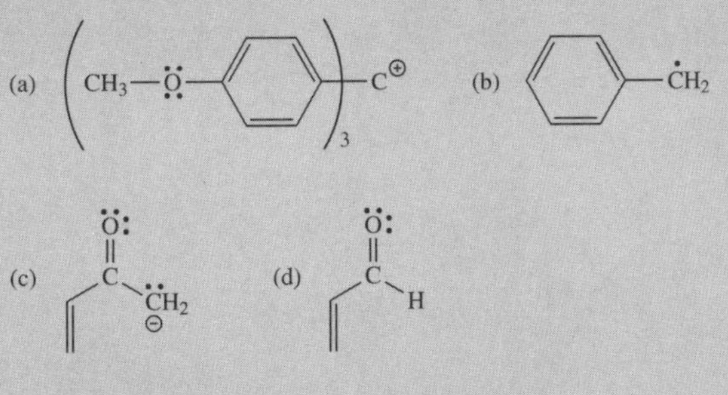

(a) $\left(CH_3-\ddot{\underset{..}{O}}-\underset{}{\bigcirc}-\overset{\oplus}{C} \right)_3$ (b) $\bigcirc-\dot{C}H_2$

(c) $\underset{\underset{\ominus}{CH_2}}{\overset{\overset{\ddot{O}:}{\parallel}}{C}}$ (d) $\underset{H}{\overset{\overset{\ddot{O}:}{\parallel}}{C}}$

1.4 Assign *R* or *S* configurations to the chiral centres in the following compounds:

[Structures shown]

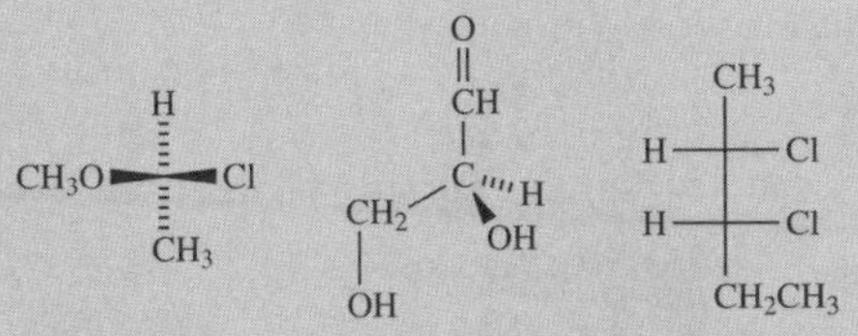

1.5 What is the total number of stereoisomers of each of the following compounds?
   (a) $(CH_3)_2CH.CHBr.CH=CH.CH_3$
   (b) $CH_3.CHOH.CHOH.CH_3$
   (c) $CH_3.CHBr.CHBr.CO_2H$
   (d) $C_2H_2Br_2$

1.6 An optically active mono methyl ester of tartaric acid (A) gives an optically inactive tartaric acid on hydrolysis. Another optically active mono methyl ester of tartaric acid (B) gives an optically active tartaric acid on hydrolysis. Suggest structures for (A) and (B).

1.7 Draw 'three-dimensional' diagrams showing the different configurations of the molecules below, and designate the chirality in each case according to the Cahn–Ingold–Prelog rules:
   $CH_3CHBrCN$ $CH_3CH_2CH(OH)NH_2$ $CH_3CH(SCH_3)CO_2CH_3$

1.8 Draw both chair conformations for *cis*-1,3-dimethylcyclohexane and *trans*-1,2-dimethylcyclohexane. Indicate in each case which conformation has the lower energy. Which conformation of *cis*-1,2-dimethylcyclohexane has the lower energy?

1.9 How might the two compounds shown be distinguished by chemical means?

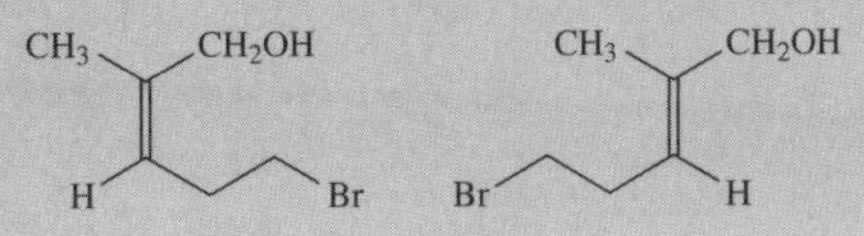

[Structures shown]

1.10 The diagram shows a plot of potential energy against dihedral angle, φ, for butane. Draw the conformations A–D.

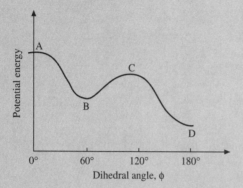

1.11 For each of the following species add the missing charge(s):

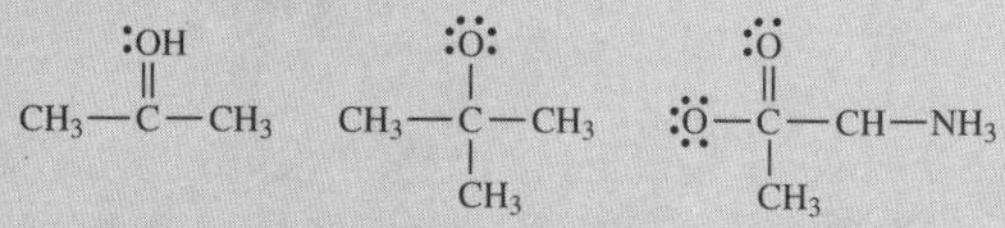

$$\underset{}{CH_3-\overset{\overset{\ddot{O}H}{\|}}{C}-CH_3} \qquad \underset{CH_3}{CH_3-\overset{\overset{:\ddot{O}:}{|}}{C}-CH_3} \qquad :\ddot{\underset{\ddot{}}{O}}-\underset{CH_3}{\overset{\overset{:\ddot{O}}{\|}}{C}-CH-NH_3}$$

1.12 For the following reactions complete the mechanistic schemes by adding the missing electrons or electron pairs, charges and curly arrows:

(a) 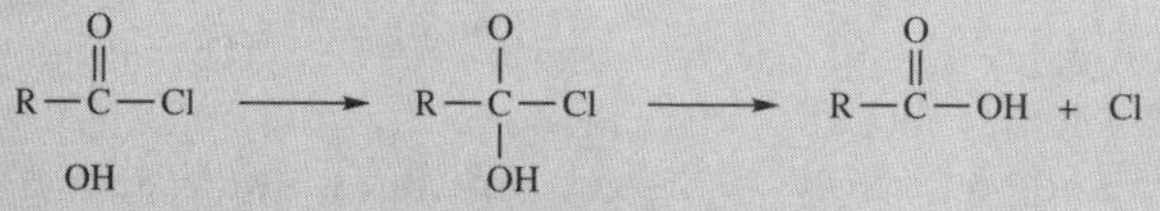 $\underset{OH}{R-\overset{\overset{O}{\|}}{C}-Cl} \longrightarrow \underset{OH}{R-\overset{\overset{O}{|}}{C}-Cl} \longrightarrow R-\overset{\overset{O}{\|}}{C}-OH + Cl$

(b)
$$RO-OR \xrightarrow{h\nu} 2RO$$
$$RO + H-Br \longrightarrow ROH + Br$$
$$Br + CH_3-CH=CH_2 \longrightarrow CH_3CH-CH_2Br$$
$$CH_3CH-CH_2Br + H-Br \longrightarrow CH_3CH_2CH_2Br + Br$$

(c) 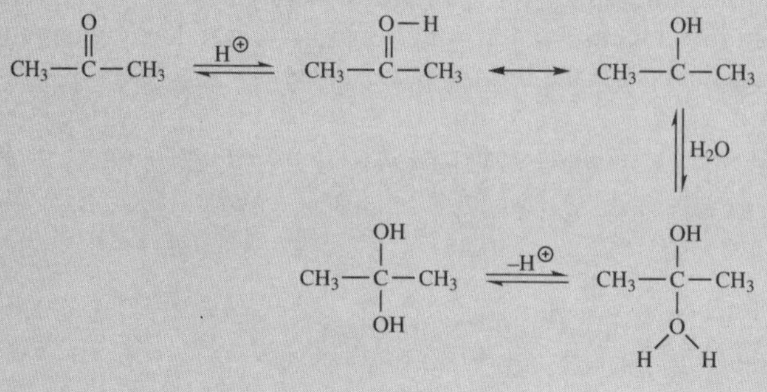 $CH_3-\overset{\overset{O}{\|}}{C}-CH_3 \underset{}{\overset{H^{\oplus}}{\rightleftharpoons}} CH_3-\overset{\overset{O-H}{\|}}{C}-CH_3 \longleftarrow CH_3-\overset{\overset{OH}{|}}{\underset{}{C}}-CH_3$

$\Updownarrow H_2O$

$\underset{OH}{CH_3-\overset{\overset{OH}{|}}{C}-CH_3} \underset{}{\overset{-H^{\oplus}}{\rightleftharpoons}} CH_3-\overset{\overset{OH}{|}}{\underset{\overset{O}{H\ \ H}}{C}}-CH_3$

1.13 By considering the mechanism for electrophilic aromatic substitution describe why anisole is activated at the *ortho* and *para* positions. (Hint: consider the number of resonance forms for attack at the *ortho*, *meta* or *para* positions.)

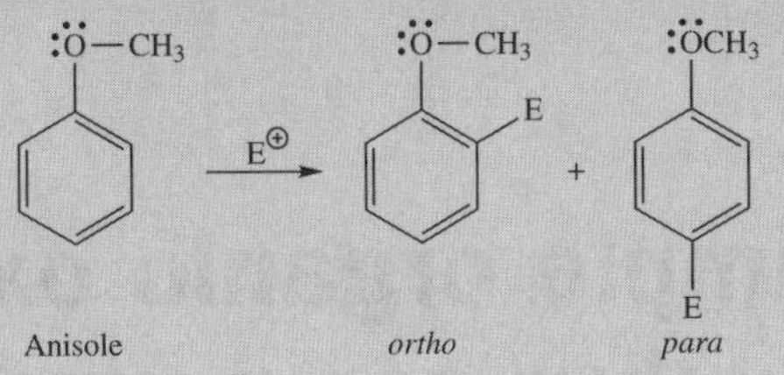

2 Simple organic oxygen and sulphur compounds: alcohols, phenols and ethers, and their sulphur analogues

Topics

2.1 Alcohols
2.2 Phenols
2.3 Ethers
2.4 Simple sulphur compounds
2.5 Summary

2.1 Alcohols

2.1.1 Nomenclature

The hydroxy derivatives of alkanes are known as alcohols, and the presence of a hydroxy group in the molecule is indicated by the prefix '**hydroxy**' or the suffix '**-ol**' in the systematic name:

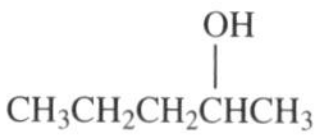

$$CH_3CH_2CH_2\overset{\overset{\displaystyle OH}{|}}{C}HCH_3$$

2-Hydroxypentane
Pentan-2-ol

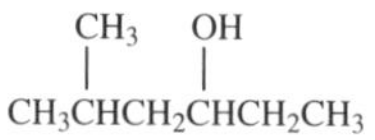

$$CH_3\overset{\overset{\displaystyle CH_3}{|}}{C}HCH_2\overset{\overset{\displaystyle OH}{|}}{C}HCH_2CH_3$$

4-Hydroxy-2-methylhexane
2-Methylhexan-4-ol

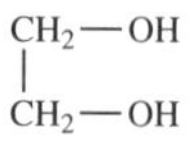

$$\begin{array}{c} CH_2-OH \\ | \\ CH_2-OH \end{array}$$

Ethane-1,2-diol
(ethylene glycol)

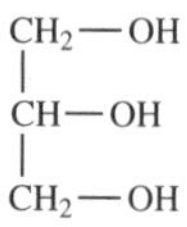

$$\begin{array}{c} CH_2-OH \\ | \\ CH-OH \\ | \\ CH_2-OH \end{array}$$

Propane-1,2,3-triol
(glycerol)

Ethane-1,2-diol and propane-1,2,3-triol are examples of polyhydric alcohols (or polyols), which are aliphatic compounds containing two or more hydroxyl groups. Most of the biologically important compounds of this type are related to carbohydrates and will be discussed later. In addition, some of the lower alcohols retain their old trivial names (Table 2.1).

Table 2.1

	Type	Systematic name	Trivial name
CH_3OH	1°	methanol	methyl alcohol
CH_3CH_2OH	1°	ethanol	ethyl alcohol
$CH_3CH_2CH_2OH$	1°	propan-1-ol	n-propyl alcohol
$CH_3CH(OH)CH_3$	2°	propan-2-ol	isopropyl alcohol
$CH_3(CH_2)_2CH_2OH$	1°	butan-1-ol	n-butyl alcohol
$CH_3CH_2CH(OH)CH_3$	2°	butan-2-ol	sec-butyl alcohol
$(CH_3)_2CHCH_2OH$	1°	2-methylpropan-1-ol	isobutyl alcohol
$(CH_3)_3COH$	3°	2-methylpropan-2-ol	t-butyl alcohol

n-, sec-, t- mean 'normal', 'secondary' and 'tertiary' respectively.

Alcohols are also named as derivatives of methanol, known as 'carbinols'. In this system, the replacement of hydrogen atoms of the methyl group is indicated as follows:

$$C_2H_5 - \underset{\underset{OH}{|}}{\overset{\overset{H}{|}}{C}} - CH_3 \qquad C_6H_5 - \underset{\underset{OH}{|}}{\overset{\overset{CH_3}{|}}{C}} - CH_3$$

Ethyl methyl carbinol   Dimethyl phenyl carbinol

Because of differences in their chemical behaviour alcohols are usually divided into three classes – primary, secondary and tertiary – which are distinguished by the number of hydrogen atoms attached to the carbon atom bearing the hydroxyl group. Primary alcohols (abbreviated by 1° or 1^y) contain the group

$$-CH_2OH$$

secondary alcohols (2° or 2^y) contain the group

$$-\overset{|}{C}HOH$$

and tertiary alcohols (3° or 3^y) which have no hydrogens on the carbon concerned, contain the group

$$-\overset{|}{\underset{|}{C}}OH$$

(see Table 2.1 for examples).

2.1.2 Preparation

It is not yet possible to introduce hydroxyl groups directly into alkanes, so all alcohol syntheses start from compounds which contain reactive functional groups.

Hydration of alkenes (olefins) in the presence of acid

$$RCH=CHR + H_2SO_4 \longrightarrow RCH_2-\underset{\underset{OSO_3H}{|}}{CHR} \xrightarrow{H_2O} RCH_2\underset{\underset{OH}{|}}{CHR}$$

This is an extremely useful reaction since alkenes are a readily available chemical feedstock – they are obtained by the cracking of alkanes (from crude oil). This is an example of an electrophilic addition to an alkene (see p. 46):

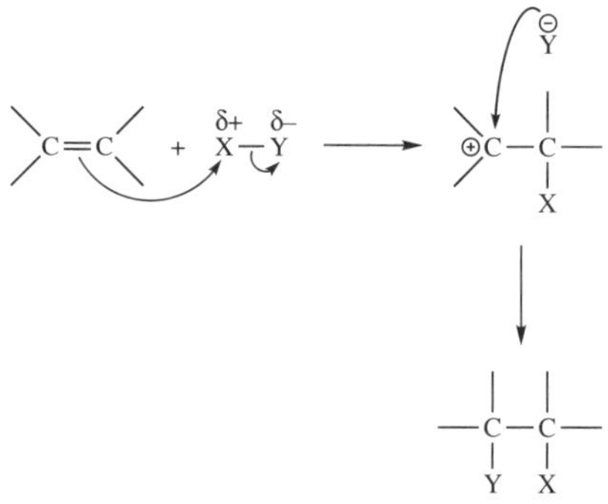

$$\underset{}{\overset{}{C=C}} + \overset{\delta+ \ \delta-}{X-Y} \longrightarrow \underset{\underset{X}{|}}{\overset{\oplus}{C}-C} \xrightarrow{Y^{\ominus}} \underset{\underset{Y \ X}{| \ |}}{-C-C-}$$

Hydrolysis of alkyl halides by water or alkali

$$\text{e.g. } CH_3CH_2CH_2-Br \xrightarrow{\overset{\ominus}{OH}/H_2O} CH_3CH_2CH_2OH + Br^{\ominus}$$

This is an example of a nucleophilic substitution, in which the nucleophile OH⁻ substitutes for the halogen (good leaving group). (See Chapter 1 for a discussion of the mechanism of these reactions.) Alkyl halides are useful intermediates in the preparation of a wide range of organic molecules since they are readily available from the addition of hydrogen halides to alkenes (see above and Chapter 1), and they undergo nucleophilic substitution by a range of nucleophiles.

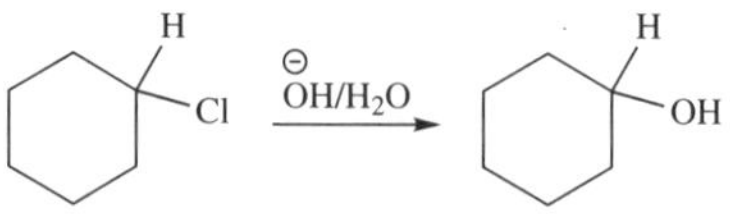

$$\text{cyclohexyl-Cl} \xrightarrow{\overset{\ominus}{OH}/H_2O} \text{cyclohexyl-OH}$$

Hydrolysis of ethers under strongly acidic conditions

$$CH_3CH_2-O-CH_3 \xrightarrow{H_2SO_4/H_2O} CH_3CH_2OH + HOCH_3$$

Alcohols

Hydrolysis of esters

1. **Acid-catalysed hydrolysis:**

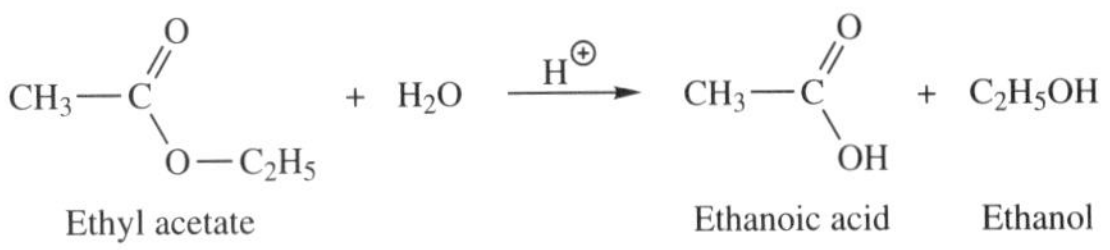

$$CH_3-\underset{O-C_2H_5}{\overset{O}{\overset{\|}{C}}} + H_2O \xrightarrow{H^{\oplus}} CH_3-\underset{OH}{\overset{O}{\overset{\|}{C}}} + C_2H_5OH$$

Ethyl acetate → Ethanoic acid + Ethanol

2. **Alkaline hydrolysis ('saponification'):**

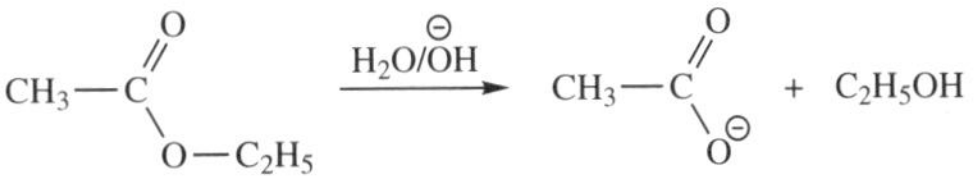

$$CH_3-\underset{O-C_2H_5}{\overset{O}{\overset{\|}{C}}} \xrightarrow{H_2O/OH^{\ominus}} CH_3-\underset{O^{\ominus}}{\overset{O}{\overset{\|}{C}}} + C_2H_5OH$$

Reduction of more highly oxidised compounds

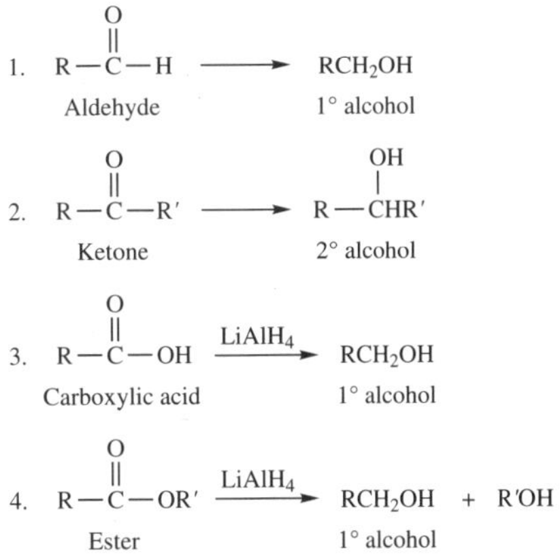

1. $R-\overset{O}{\overset{\|}{C}}-H \longrightarrow RCH_2OH$
   Aldehyde → 1° alcohol

2. $R-\overset{O}{\overset{\|}{C}}-R' \longrightarrow R-\underset{}{\overset{OH}{\overset{|}{C}}}HR'$
   Ketone → 2° alcohol

3. $R-\overset{O}{\overset{\|}{C}}-OH \xrightarrow{LiAlH_4} RCH_2OH$
   Carboxylic acid → 1° alcohol

4. $R-\overset{O}{\overset{\|}{C}}-OR' \xrightarrow{LiAlH_4} RCH_2OH + R'OH$
   Ester → 1° alcohol

The reduction of aldehydes and ketones can be accomplished by a variety of reducing agents, including catalytic reduction (hydrogen and nickel or platinum catalyst), lithium aluminium hydride ($LiAlH_4$) or sodium borohydride ($NaBH_4$), or 'dissolving metal'* reducing agents (zinc and hydrochloric or acetic acid, sodium amalgam and water, or sodium and ethanol). Carboxylic acids and their esters, however, can only be reduced by lithium aluminium hydride.

The Grignard synthesis of alcohols

Grignard reagents, which are conveniently prepared from an alkyl halide, RX, and magnesium under anhydrous conditions, are carbon nucleophiles and so react with aldehydes and ketones to form a bond between the carbon bound to magnesium and the carbonyl carbon.

---

*Dissolving metal reductions employ a reactive metal and an acid. The reduction process consists of electron transfer from the metal to the substrate, along with proton capture from the acid present.

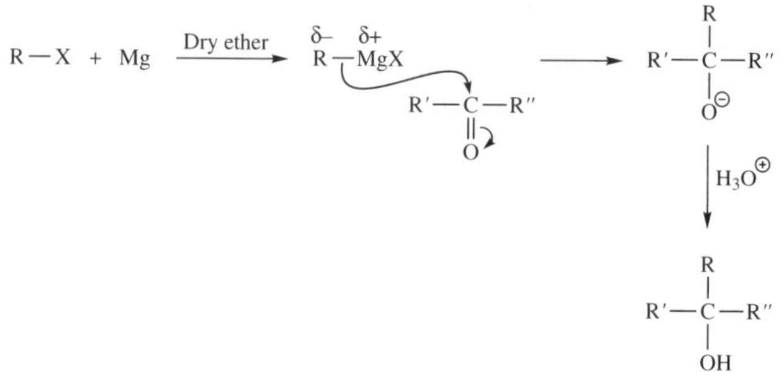

$$R-X + Mg \xrightarrow{\text{Dry ether}} \overset{\delta-}{R}-\overset{\delta+}{MgX} \longrightarrow R'-\underset{\underset{O^\ominus}{|}}{\overset{\overset{R}{|}}{C}}-R''$$

$$R'-\underset{\|}{\overset{}{C}}-R''$$
$$O$$

$$\Big\downarrow H_3O^\oplus$$

$$R'-\underset{\underset{OH}{|}}{\overset{\overset{R}{|}}{C}}-R''$$

This is a classical method of preparing alcohols since a variety of alcohols can be prepared, depending upon the alkyl halide and the carbonyl compound used (p. 94).

Reaction of primary amines with nitrous acid

$$\text{e.g. } R-CH_2-NH_2 \xrightarrow[H_2O]{NaNO_2/HCl} R-CH_2OH$$

This reaction is seldom used preparatively, owing to the number of by-products formed.

'Biological' preparations of alcohols

Ethanol, C_2H_5OH, can be obtained by the fermentation of sugar (glucose) by yeast:

$$C_6H_{12}O_6 \xrightarrow{\text{Yeast}} C_2H_5OH$$

Glycerol (propane-1,2,3-triol) can also be produced by the fermentation of glucose under special conditions but is more usually obtained by the hydrolysis of animal fats or plant oils, which are naturally occurring esters of glycerol and long chain carboxylic acids:

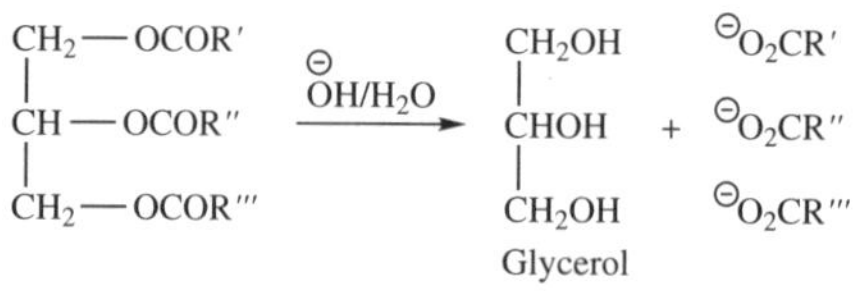

$$\begin{array}{c} CH_2-OCOR' \\ | \\ CH-OCOR'' \\ | \\ CH_2-OCOR''' \end{array} \xrightarrow{{}^\ominus OH/H_2O} \begin{array}{c} CH_2OH \\ | \\ CHOH \\ | \\ CH_2OH \\ \text{Glycerol} \end{array} + \begin{array}{c} {}^\ominus O_2CR' \\ {}^\ominus O_2CR'' \\ {}^\ominus O_2CR''' \end{array}$$

Preparation of *vicinal*-diols

Diols are readily prepared by direct hydroxylation of alkenes using dilute potassium permanganate:

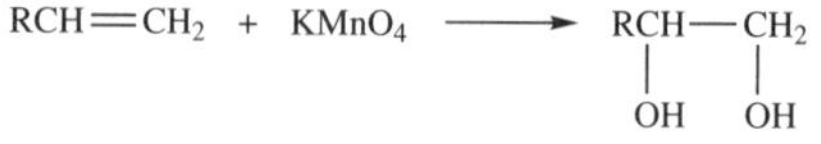

$$RCH=CH_2 + KMnO_4 \longrightarrow \underset{\underset{OH}{|}}{RCH}-\underset{\underset{OH}{|}}{CH_2}$$

This can also be achieved by reaction of the alkene with osmium tetroxide, and decomposition of the addition compound with sodium sulphite:

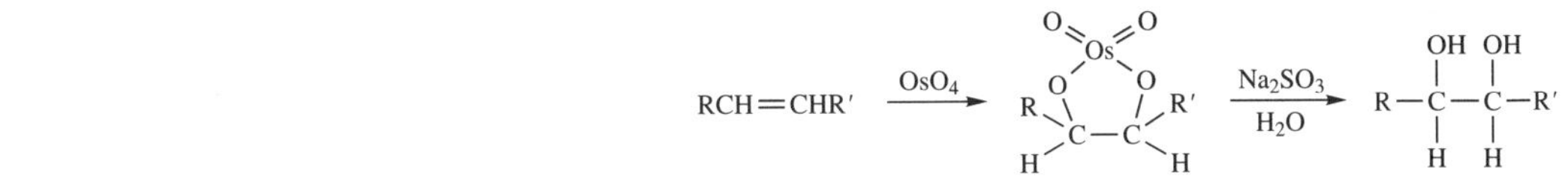

$$RCH=CHR' \xrightarrow{OsO_4} \underset{\underset{H}{C}-\underset{H}{C}}{\overset{O\diagdown\underset{\diagup O}{Os}\diagup O}{R\diagup\diagdown R'}} \xrightarrow[H_2O]{Na_2SO_3} R-\underset{\underset{H}{|}}{\overset{\overset{OH}{|}}{C}}-\underset{\underset{H}{|}}{\overset{\overset{OH}{|}}{C}}-R'$$

Other methods of preparing *vic*-diols (1,2-diols) are shown below for the preparation of ethane-1,2-diol (which, when mixed with water, is used as 'antifreeze').

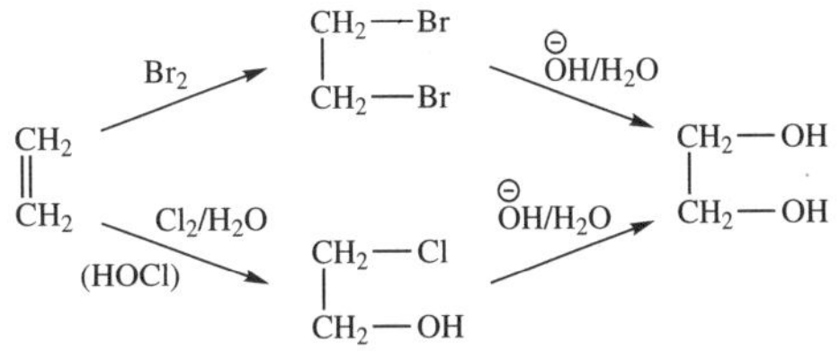

$$CH_2{=}CH_2 \begin{array}{c} \xrightarrow{Br_2} \\ \\ \xrightarrow[(HOCl)]{Cl_2/H_2O} \end{array} \begin{array}{c} CH_2{-}Br \\ | \\ CH_2{-}Br \\ \\ CH_2{-}Cl \\ | \\ CH_2{-}OH \end{array} \begin{array}{c} \xrightarrow{{}^{\ominus}OH/H_2O} \\ \\ \xrightarrow{{}^{\ominus}OH/H_2O} \end{array} \begin{array}{c} CH_2{-}OH \\ | \\ CH_2{-}OH \end{array}$$

2.1.3 Reactions

Salt formation

Although alcohols are neutral to indicators and are virtually undissociated in aqueous solution, the hydrogen of the hydroxyl group can be removed by the direct reaction of an alcohol with the electropositive metals of groups I and II of the periodic table:

$$C_2H_5OH + Na \longrightarrow C_2H_5O^{\ominus} Na^{\oplus} + H_2$$
Sodium ethoxide

$$C_2H_5OH + Mg \longrightarrow (C_2H_5O^{\ominus})_2Mg + H_2$$
Magnesium ethoxide

Since alcohols are extremely weak acids, their salts – the alkoxides – are very strong bases. The basicity of alkoxides depends on the class of the parent alcohol, the tertiary alkoxides (e.g. $(CH_3)_3CO^-$) being the strongest bases, and methoxide the weakest, i.e. basicity $R_3CO^- > R_2CHO^- > RCH_2O^- > CH_3O^-$ (R = alkyl group). The alkoxides are also good nucleophiles, reacting with alkyl halides to form ethers (p. 71).

Ethane-1,2-diol reacts with sodium to give a monosodium salt, $HOCH_2-CH_2O^-Na^+$. Further reaction of the second hydroxyl group is difficult, since this would produce a second negative charge on the anion, close to that already present. Glycerol reacts with sodium to form a monosodium salt, in which two alkoxide anions are in equilibrium, leading to mixtures of products from reactions of the sodium salts:

64 Simple organic oxygen and sulphur compounds

$$\text{Glycerol: } \underset{\substack{\text{CH}_2\text{OH} \\ | \\ \text{CHOH} \\ | \\ \text{CH}_2\text{OH}}}{} \xrightarrow{\text{Na}} \left\{ \begin{array}{c} \underset{\substack{\text{CH}_2\text{O}^\ominus\text{Na}^\oplus \\ | \\ \text{CHOH} \\ | \\ \text{CH}_2\text{OH}}}{} \\ \updownarrow \\ \underset{\substack{\text{CH}_2\text{OH} \\ | \\ \text{CHO}^\ominus\text{Na}^\oplus \\ | \\ \text{CH}_2\text{OH}}}{} \end{array} \right\} \xrightarrow{\text{CH}_3\text{I}} \underset{\substack{\text{CH}_2\text{OCH}_3 \\ | \\ \text{CHOH} \\ | \\ \text{CH}_2\text{OH}}}{} + \underset{\substack{\text{CH}_2\text{OH} \\ | \\ \text{CHOCH}_3 \\ | \\ \text{CH}_2\text{OH}}}{}$$

Ester formation

Alcohols react with carboxylic acids in the presence of mineral acid catalysts, to give esters:

$$\text{R}-\underset{\text{OH}}{\overset{\text{O}}{\text{C}}} + \text{CH}_3\text{OH} \xrightleftharpoons{\text{H}^\oplus} \text{R}-\underset{\text{O}-\text{CH}_3}{\overset{\text{O}}{\text{C}}} + \text{H}_2\text{O}$$

For the preparation of esters, however, it is frequently better to react the alcohol with either an acyl chloride or acid anhydride (p. 112):

$$\underset{\substack{\text{OH} \\ | \\ \text{CH}_3\text{CHCH}_3}}{} + \left\{ \begin{array}{c} \text{CH}_3\overset{\text{O}}{\overset{\|}{\text{C}}}\text{Cl} \\ \text{Acetyl chloride} \\ \textit{or} \\ \text{CH}_3-\overset{\text{O}}{\overset{\|}{\text{C}}}-\text{O}-\overset{\text{O}}{\overset{\|}{\text{C}}}-\text{CH}_3 \\ \text{Acetic anhydride} \end{array} \right\} \longrightarrow \underset{\substack{\text{O}-\overset{\text{O}}{\overset{\|}{\text{C}}}-\text{CH}_3 \\ | \\ \text{CH}_3\text{CHCH}_3}}{} + \left\{ \begin{array}{c} \text{HCl} \\ \textit{or} \\ \text{CH}_3\text{CO}_2\text{H} \end{array} \right.$$

Ethane-1,2-diol reacts in most cases as a typical alcohol, in which the two similar functional groups can behave independently. This leads to a more complex pattern of derivatives than is found in simple alcohols, and two series of esters, ethers, halides, etc., can be formed from the diol:

$$\underset{\substack{\text{CH}_2-\text{OH} \\ | \\ \text{CH}_2-\text{OH}}}{} \xrightarrow{\text{CH}_3\text{CO}_2\text{H}/\text{H}^\oplus} \underset{\substack{\text{CH}_2-\text{O}-\overset{\text{O}}{\overset{\|}{\text{C}}}-\text{CH}_3 \\ | \\ \text{CH}_2-\text{OH}}}{} \xrightarrow{\text{CH}_3\text{CO}_2\text{H}/\text{H}^\oplus} \underset{\substack{\text{CH}_2-\text{O}-\overset{\text{O}}{\overset{\|}{\text{C}}}-\text{CH}_3 \\ | \\ \text{CH}_2-\text{O}-\overset{}{\overset{}{\text{C}}}-\text{CH}_3 \\ \phantom{\text{CH}_2-\text{O}-}\overset{\|}{\text{O}}}}{}$$

$$\downarrow \text{C}_6\text{H}_5\text{CO}_2\text{H}/\text{H}^\oplus$$

$$\underset{\substack{\text{CH}_2-\text{O}-\overset{\text{O}}{\overset{\|}{\text{C}}}-\text{CH}_3 \\ | \\ \text{CH}_2-\text{O}-\overset{}{\overset{}{\text{C}}}-\text{C}_6\text{H}_5 \\ \phantom{\text{CH}_2-\text{O}-}\overset{\|}{\text{O}}}}{}$$

(In the general case of a diol in which the two hydroxyl groups are not identical, e.g. CH$_3$CH(OH)CH$_2$OH, two different monoderivatives can be formed.) The reactions of glycerol are those expected of a compound that is both a primary and secondary alcohol and several series of derivatives can be prepared.

Oxidation

The oxidation of alcohols (opposite of the reduction of aldehydes/ketones) gives products which vary according to the class of alcohol.

Primary (1°) alcohols can be oxidised to aldehydes, which can be further oxidised to carboxylic acids. The use of mild oxidants, such as chromic acid (Na$_2$Cr$_2$O$_7$ + H$_2$SO$_4$; or CrO$_3$), or, better still, pyridinium chlorochromate (PCC), allows the isolation of the aldehyde (p. 81). Powerful oxidising agents, such as potassium permanganate or concentrated nitric acid, give only the carboxylic acids:

$$R-CH_2OH \; (1°) \xrightarrow[\text{(PCC)}]{CrO_3 \text{ or } CrO_3Cl^{\ominus} / \text{pyridine}} R-CHO \text{ (Aldehyde)} \xrightarrow{CrO_3} R-CO_2H \text{ (Carboxylic acid)}$$

$$R-CH_2OH \xrightarrow{HNO_3 \text{ or } KMnO_4} R-CO_2H$$

Secondary alcohols are oxidised readily to ketones, which are much more resistant to oxidation (p. 83).

$$R-\underset{\underset{(2°)}{|}}{\overset{\overset{OH}{|}}{CH}}-R' \xrightarrow{CrO_3} R-\overset{\overset{O}{\|}}{C}-R' \text{ Ketone}$$

Tertiary alcohols are resistant to oxidation under mild conditions; under more vigorous conditions, decomposition of the carbon skeleton takes place via C—C bond cleavage.

The oxidation of ethane-1,2-diol (e.g. by HNO$_3$ or CrO$_3$) can lead to five possible products, depending upon the sequence and extent of oxidation of the two primary alcohol groups present. Only the final product, oxalic acid, is readily obtained by this method, partial oxidation leading to mixtures of intermediates:

$$\begin{array}{c} CH_2-OH \\ | \\ CH_2-OH \end{array} \xrightarrow{[O]} \begin{array}{c} CH=O \\ | \\ CH_2OH \end{array} \text{Glycolaldehyde}$$

then branching via [O] to:

- Glyoxal: CH=O / CH=O
- Glycollic acid: O=C(OH) / CH$_2$OH

then to Glyoxylic acid: CO$_2$H / CH=O

$$\xrightarrow{[O]} \begin{array}{c} CO_2H \\ | \\ CO_2H \end{array} \text{Oxalic acid (ethanedioic acid)}$$

The oxidation of glycerol (propane-1,2,3-triol) with powerful oxidants, such as chromic acid, leads to complete degradation of the molecule. Milder oxidising agents (dilute nitric acid or sodium hypobromite) give an equilibrium (tautomeric; see p. 91) mixture of the expected aldehyde and ketone, and further oxidation with nitric acid gives glyceric acid:

$$\begin{array}{c} CH_2OH \\ | \\ CHOH \\ | \\ CH_2OH \end{array} \xrightarrow{NaOBr} \begin{array}{c} CH_2OH \\ | \\ C=O \\ | \\ CH_2OH \end{array} \rightleftharpoons \begin{array}{c} CH=O \\ | \\ CHOH \\ | \\ CH_2OH \end{array} \xrightarrow{HNO_3} \begin{array}{c} CO_2H \\ | \\ CHOH \\ | \\ CH_2OH \end{array}$$

$$\text{Dihydroxypropanone} \qquad \text{Glyceraldehyde} \qquad \text{Glyceric acid}$$

Reaction with sulphuric acid

Alcohols react with sulphuric acid in three ways. Under mild conditions alkyl hydrogen sulphates are formed:

$$ROH + H_2SO_4 \xrightarrow{80°C} R-O-SO_3H + H_2O$$
$$\text{Alkyl hydrogen sulphate}$$

while under more vigorous conditions, dehydration to ethers (p. 71) or alkenes occurs. Tertiary alcohols are particularly easily dehydrated to alkenes.

$$CH_3-\underset{\underset{OH}{|}}{CH}-CH_3 \xrightarrow[\substack{H_3PO_4 \\ (-H_2O)}]{H_2SO_4 \text{ or}} CH_3-CH=CH_2$$

Alkyl halide formation

Reaction with hydrogen halides. Alcohols react with hydrogen halides to give alkyl halides:

$$R-OH + HBr \longrightarrow R-Br + H_2O$$

This reaction proceeds in two steps. Initial protonation of the hydroxyl group by the hydrogen halide forms an 'oxonium ion', which is attacked by the halide anion, resulting in the elimination of water and formation of the alkyl halide:

$$R-\ddot{O}-H \xrightarrow{H^{\oplus}} R-\overset{\overset{H}{|}}{\underset{\underset{:\ddot{B}r:^{\ominus}}{}}{\overset{\oplus}{O}}}-H \longrightarrow R-Br + H_2O$$

e.g.

$$CH_3CH_2-\underset{\underset{OH}{|}}{CH}-CH_3 + HI \longrightarrow CH_3CH_2-\underset{\underset{I}{|}}{CH}CH_3 + H_2O$$

Reaction with non-metal halides. Alcohols also react with a number of non-metal halides to give alkyl halides:

$$\text{ROH} + \begin{cases} \text{PCl}_3 \\ \text{or} \\ \text{PCl}_5 \\ \text{or} \\ \text{SOCl}_2 \end{cases} \longrightarrow \text{R}-\text{Cl} + \begin{cases} \text{H}_3\text{PO}_3 \\ \text{or} \\ \text{HCl} + \text{POCl}_3 \\ \text{or} \\ \text{SO}_2 + \text{HCl} \end{cases}$$

The method of choice usually involves thionyl chloride ($SOCl_2$) since both by-products are gaseous.

Oxidation of polyols

Polyols which have hydroxyl groups on adjacent carbon atoms (*vicinal* or *vic*-diols) are oxidised by periodic acid, HIO_4, or lead tetra-acetate, $Pb(OCOCH_3)_4$, with cleavage of the bond between the hydroxyl-substituted carbon atoms, leading to the production of carbonyl compounds:

$$\begin{array}{c} CH_2OH \\ | \\ CH_2OH \end{array} \xrightarrow[\text{or } Pb(OCOCH_3)_4]{HIO_4} \begin{array}{c} CH_2=O \\ + \\ CH_2=O \end{array} \qquad \begin{array}{c} CH_2OH \\ | \\ CHOH \\ | \\ CH_2OH \end{array} \xrightarrow{2HIO_4} \begin{array}{c} CH_2=O \\ + \\ HCO_2H \\ + \\ CH_2=O \end{array}$$

Periodic acid is reduced to iodic acid, HIO_3, in the process, and lead tetra-acetate to lead(II) acetate and acetic acid. These reactions are quantitative and of great use in the study of polyols containing a sequence of *vic*-diols. Oxidation with either reagent leads to quantitative degradation to characteristic fragments:

$$\begin{array}{c} CH_2OH \\ | \\ CHOH \\ | \\ R-C-OH \\ | \\ CHOH \\ | \\ R' \end{array} \xrightarrow{3HIO_4} \begin{array}{c} CH_2=O \\ + \\ HCO_2H \\ + \\ RCO_2H \\ + \\ R'CH=O \end{array} + 3HIO_3$$

2.2 Phenols

Aromatic hydroxy-compounds in which the hydroxyl group is attached to an aromatic ring are called phenols, and the name 'phenol' is also applied to the simplest compound of the series, hydroxybenzene (C_6H_5OH). Typical examples of phenols are:

4-Hydroxymethylbenzene (*p*-cresol)

1,2-Dihydroxybenzene (catechol)

2-Chloro-5-hydroxyethylbenzene

2.2.1 Preparation

From aromatic primary amines (via diazonium salts)
Aromatic primary amines react with nitrous acid to form diazonium salts and, if these are allowed to warm to above 5 °C, they form phenols (p. 132):

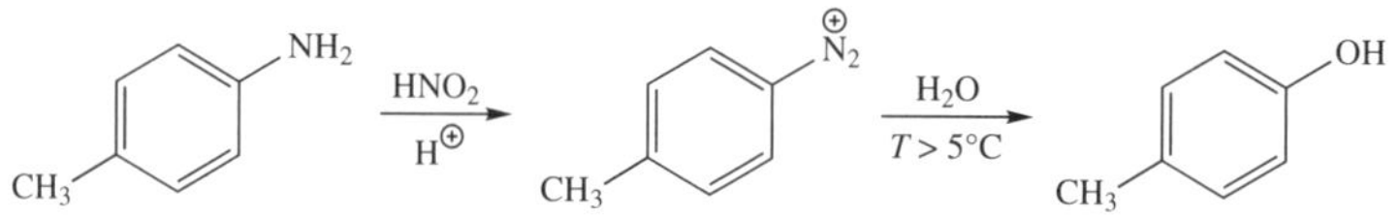

[Reaction scheme: 4-methylaniline (CH$_3$-C$_6$H$_4$-NH$_2$) $\xrightarrow{\text{HNO}_2, \text{H}^\oplus}$ 4-methylbenzenediazonium (CH$_3$-C$_6$H$_4$-N$_2^\oplus$) $\xrightarrow{\text{H}_2\text{O},\ T>5°C}$ 4-methylphenol (CH$_3$-C$_6$H$_4$-OH)]

From aromatic sulphonic acids
Aromatic sulphonic acids react with fused alkali hydroxides to give phenols:

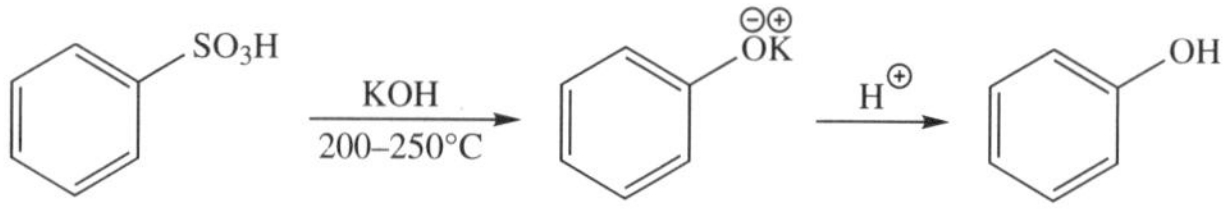

[Reaction scheme: C$_6$H$_5$-SO$_3$H $\xrightarrow{\text{KOH},\ 200-250°C}$ C$_6$H$_5$-OK $\xrightarrow{\text{H}^\oplus}$ C$_6$H$_5$-OH]

Aryl halides are generally inert to nucleophilic substitution reactions which are thus not normally possible except under very vigorous conditions. Chlorobenzene can only be hydrolysed with aqueous alkali above 300 °C under high pressure, and this is the method used industrially for the preparation of phenol itself (Dow process).

2.2.2 Reactions

The reactions of phenols can be classified into two types: reactions of the —OH group and reactions of the aromatic ring.

Reactions of the —OH group
The most characteristic property of phenols is the feeble acidity of the hydroxyl group, and phenols will readily dissolve in dilute aqueous sodium hydroxide, producing the phenoxide anion:

[Equilibrium: 3-methylphenol + $^\ominus$OH $\rightleftharpoons$ 3-methylphenoxide ($^\ominus$O-C$_6$H$_4$-CH$_3$) + H$_2$O]

Phenol itself has a pK_a of 9.8 (p. 107) and simple alkyl-substituted phenols have comparable acidities. Carbonic acid, H$_2$CO$_3$ (pK_1 = 6.56), is approximately a thousand times more strongly acidic, so most simple phenols do not react with sodium bicarbonate solution, and can be precipitated from solutions of the phenoxide by saturation with carbon dioxide:

$$\text{Ar}-\text{O}^\ominus + \text{CO}_2 + \text{H}_2\text{O} \longrightarrow \text{Ar}-\text{OH} + \text{HCO}_3^\ominus$$
(Ar = aryl group)

The acidity of phenols is attributable to resonance stabilisation of the anion, for which four canonical structures (p. 13) are possible:

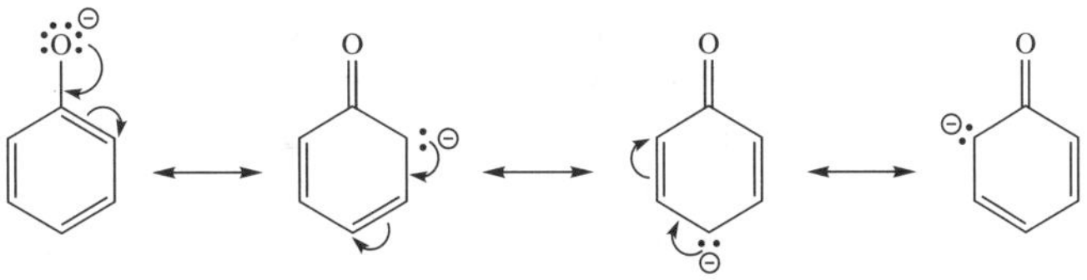

The effect of this resonance is to distribute the negative charge of the anion over the molecule, rather than leaving it concentrated on one particular atom, as in the case of the alkoxide anions. The consequent decrease in the energy required to form a phenoxide anion shows itself in the ease of ionisation of the phenol, i.e. its greater acidity, compared with alcohols (cf. the ease of formation of carbocations, p. 44).

The acidity of a phenol is greatly increased by the presence of powerful electronegative substituents on the aromatic ring in positions *ortho-* or *para-* to the hydroxyl group.

Few of the reactions of alcoholic hydroxyl groups occur with phenols. Esters are readily formed by reaction with acid anhydrides or chlorides, and direct esterification with carboxylic acids is also possible:

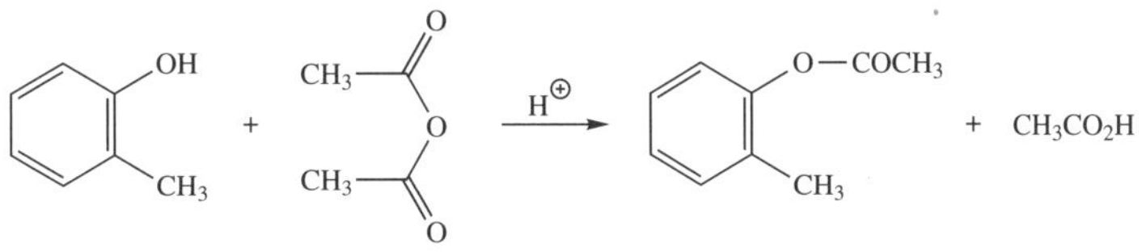

The direct formation of ethers by the action of sulphuric acid is not possible, but preparation of alkyl aryl ethers via the phenoxide anion occurs normally:

Oxidation of phenols occurs readily, but leads to complex products, often with simultaneous formation of much tarry material. In suitable circumstances, quinones (p. 95) are formed.

Phenols cannot normally be converted directly into the corresponding aryl halides. Hydrogen halides have no effect, and reaction with phosphorus halides leads predominantly to the aryl esters of phosphorus oxyacids. Phenols, like many other compounds containing the 'enol' group,

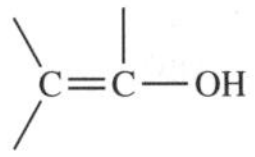

$$\diagdown_{\diagup}C=C-OH$$

form intensely coloured complexes with ferric ion in neutral solution. The blue, purple or green colours produced are often used as a qualitative test for phenols, but are also given by certain aliphatic compounds, which exist in solution partly as the enol (p. 91).

Reactions of the aromatic ring

Phenols are very readily attacked by electrophiles, with substitution occurring in the *ortho-* and *para-* positions. It is often difficult to prevent substitution occurring more than once. Dilute nitric acid rapidly converts phenol into its *ortho-* and *para-*nitro derivatives, while under the conditions in which benzene is converted into nitrobenzene, phenol is nitrated three times, forming picric acid (2,4,6-trinitrophenol). In aqueous solution, chlorine or bromine water will give the corresponding trisubstituted derivatives. Even the feebly electrophilic nitrosonium ion, NO^+, produced in acidified nitrous acid solution, will convert phenol into its *p*-nitroso derivative.

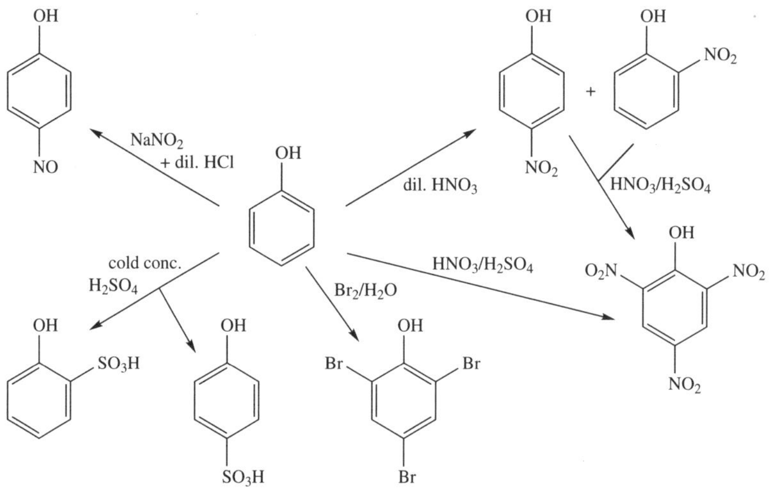

2.3 Ethers

Ethers may be regarded as the dialkyl or diaryl, etc., derivatives of water (making the alcohols the mono-alkyl derivatives), e.g.

$$\begin{array}{ccc} H-O-H & R-O-H & R-O-R \\ \text{Water} & \text{Alcohol} & \text{Ether} \end{array}$$

Although ether linkages (i.e. C—O—C) are frequently found in naturally occurring compounds, simple ethers are of little biological significance. Thyroxine – the thyroid hormone – is a naturally occurring diaryl ether:

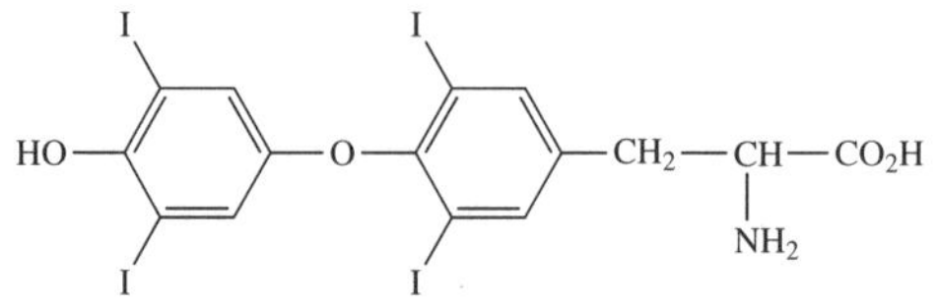

2.3.1 Preparation

Dehydration of alcohols

Direct dehydration of alcohols by sulphuric acid is possible with primary alcohols, but secondary and tertiary alcohols are too readily converted into alkenes (p. 51). Initial conversion of part of the primary alcohol into the alkyl hydrogen sulphate is followed by alkylation of the residual alcohol by this ester of sulphuric acid:

$$C_2H_5OH \xrightarrow[80°C]{H_2SO_4} C_2H_5-OSO_3H \xrightarrow[150°C]{C_2H_5OH} (C_2H_5)_2O + H_2SO_4$$

Williamson's synthesis

A more general preparation of ethers is the reaction of nucleophilic alkoxide anions with alkyl halides (Williamson's synthesis):

$$R'-\ddot{O}:^{\ominus} \quad R-\ddot{B}r: \longrightarrow R'-\ddot{O}-R + :\ddot{B}r:^{\ominus}$$

Since the alkoxide and the alkyl halide need not have the same alkyl group, this method is suitable for the preparation of mixed ethers:

$$C_6H_5-O^{\ominus} + CH_3Br \longrightarrow C_6H_5-O-CH_3 + Br^{\ominus}$$

Methyl phenyl ether (anisole)

$$CH_3O^{\ominus} + CH_3CH_2CH_2CH_2Br \longrightarrow CH_3CH_2CH_2CH_2OCH_3 + Br^{\ominus}$$

Butyl methyl ether

2.3.2 Reactions

Ethers are relatively inert and thus are widely used as solvents in organic reactions (they do not react with the reagents). They do not react with most of the reagents which attack alcohols, as they lack the chemically reactive hydroxyl group. They are completely inert to alkalis or alkali metals, but under strongly acidic conditions they are converted into oxonium cations by protonation of the oxygen atom, and these cations may react with nucleophiles. Thus, hydrogen bromide or iodide cleaves the ether link (cf. the mechanism of conversion of alcohols into alkyl halides, p. 66):

$$R-\ddot{O}-R \quad H^{\oplus} \rightleftharpoons R-\overset{\overset{H}{|}}{\underset{}{\overset{\oplus}{O}}}-R \longrightarrow R-I + R-\ddot{O}H$$
$$:\ddot{I}:^{\ominus}$$

Oxonium cation

$$[ROH + HI \longrightarrow RI + H_2O]$$

Electrophilic substitution of the aromatic ring of aryl ethers, e.g. anisole, occurs predominantly *ortho-* and *para-* with respect to the alkoxy group:

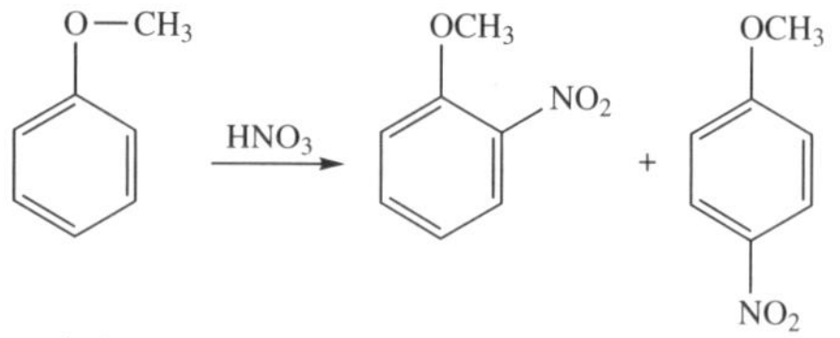

2.4 Simple sulphur compounds

The replacement of the hydrogen atoms of water by alkyl or aryl groups leads to alcohols or phenols, and ethers. In precisely the same way, the hydrides of sulphur, H_2S and H_2S_2, are the parents of three types of simple aliphatic sulphur compounds:

R—S—H   R—S—R   R—S—S—R
Thiols   Thioethers   Disulphides
(mercaptans)

Aliphatic sulphur compounds are widely distributed in nature, and are of great biological importance. The protein chains of enzymes frequently contain thiol groups, which are vital for the catalytic activity of the enzyme, and the poisonous properties of some heavy metals, e.g. arsenic, lead and mercury, are due to their ability to combine with these thiol groups, thereby interfering with cell reactions. These enzymes are also inhibited by treatment with iodoethanoic acid, ICH_2CO_2H, a powerful alkylating agent which converts the —SH group into —SCH_2CO_2H. Many carboxylic acids are utilised by cells in the form of their esters of coenzyme A, a complex nucleotide thiol.

Sulphonium salts are known to be intermediates in some biological alkylations. Disulphide groups are important structural features in many proteins and polypeptide hormones, e.g. insulin, and certain reduction–oxidation reactions of cells utilise thiol–disulphide interconversion as a redox system (see below). The evil-smelling secretion of skunks contains much 3-methylbutane-1-thiol, $(CH_3)_2CHCH_2CH_2SH$, with *trans*-but-2-ene-1-thiol, $CH_3CH=CHCH_2SH$, and the disulphide derivative $CH_3CH=CHCH_2SSCH_3$. Sulphur compounds are also responsible for the characteristic odours of *Allium* species (onions and garlic); garlic oil contains much diallyl disulphide, $CH_2=CHCH_2(-S-)_n-CH_2CH=CH_2$ ($n = 2$), with the corresponding tri- and tetrasulphides ($n = 3, 4$), and also a recently identified potent antithrombotic compound 'ajoene'. The lachrymatory component of onion juice has been identified as the *S*-oxide of thiopropanal.

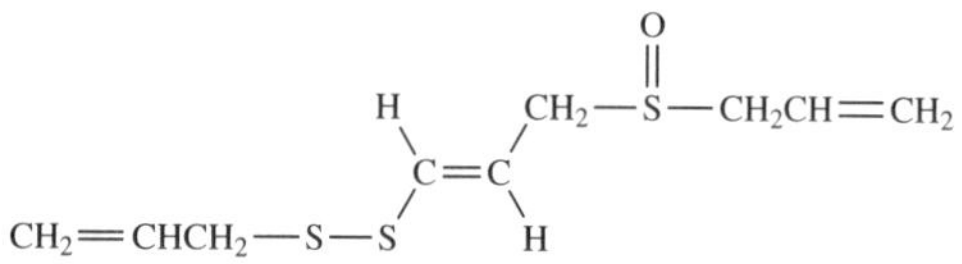

Ajoene

$$\underset{C_2H_5}{\overset{H}{>}}C=\overset{\oplus}{S}\overset{}{\underset{O^{\ominus}}{>}}\quad \text{Thiopropanal-}S\text{-oxide}$$

Thioctic acid (*α-lipoic acid*) is a naturally occurring disulphide, which is a cofactor required for the enzymic oxidation of pyruvic acid to acetic acid. The disulphide is the oxidising agent, being reduced to the corresponding thiol, which can subsequently be reoxidised.

$$\underset{\text{Thioctic acid}}{\overset{S-S}{\underset{H\ H}{H_2C\diagdown C\diagup CH-(CH_2)_4CO_2H}}} \underset{-2H-2\bar{e}\ (\text{oxidation})}{\overset{+2H+2\bar{e}\ (\text{reduction})}{\rightleftarrows}} \underset{\text{6,8-Dimercaptooctanoic acid}}{\overset{HS\quad SH}{\underset{CH_2}{CH_2\diagdown \diagup CH(CH_2)_4CO_2H}}}$$

The overall reaction for the conversion of pyruvic acid into the ester of acetic acid and coenzyme A (acetyl CoA) is:

$$\underset{\text{Pyruvic acid}}{CH_3-\overset{O}{\overset{\|}{C}}-C\overset{O}{\underset{OH}{\diagup\!\!\diagdown}}} + \underset{\text{Coenzyme A}}{CoA-SH} + \overset{S-S}{\underset{CH_2}{H_2C\diagdown\diagup CH(CH_2)_4CO_2H}}$$

$$\downarrow + \text{enzyme}$$

$$\underset{\substack{\text{Acetylcoenzyme A}\\\text{Ethanoylcoenzyme A}}}{CH_3-\overset{O}{\overset{\|}{C}}-S-CoA} + CO_2 + \overset{SH\quad SH}{\underset{CH_2}{H_2C\diagdown\diagup CH(CH_2)_4CO_2H}}$$

2.4.1 Preparation

Thiols may be prepared by the reaction of alkali hydrosulphides with alkyl halides, or by the action of phosphorus sulphides on alcohols:

$$H-\overset{..}{\underset{..}{S}}{}^{\ominus} \curvearrowright R\overset{..}{\underset{..}{\frown}}\overset{..}{\underset{..}{I}}: \longrightarrow H-\overset{..}{\underset{..}{S}}-R + :\overset{..}{\underset{..}{I}}:{}^{\ominus}$$

$$R-OH + P_2S_5 \longrightarrow R-SH$$

Thioethers may be obtained by the action of alkali sulphides on alkyl halides, or by the alkylation of thiolate anions by alkyl halides (cf. Williamson's ether synthesis, p. 71):

$$R-Br + S^{2-} \longrightarrow R-S-R \quad (\text{in two steps})$$

$$RS^{\ominus} + R'-I \longrightarrow R-S-R' + I^{\ominus}$$

Dialkyl disulphides are produced by the oxidation of the corresponding thiols, into which they may be converted by reducing agents:

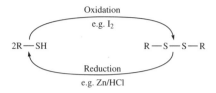

2.4.2 Reactions

Although sulphur is less electronegative than oxygen, *thiols* are much more strongly acidic than alcohols (CH_3CH_2SH, $pK = 11$) and can form salts, 'thiolates' or 'mercaptides', corresponding to the alkoxides, even in aqueous alkali:

$$R-SH + \overset{\ominus}{O}H \rightleftharpoons RS^{\ominus} + H_2O$$

Covalent heavy metal derivatives, e.g. $Hg(SC_2H_5)_2$ can also be precipitated from aqueous solutions.

Thiols, like alcohols, can be esterified by carboxylic acids, acyl chlorides or acid anhydrides:

$$R-SH + R'-CO_2H \xrightleftharpoons{H^{\oplus}} R-S-CO-R' + H_2O$$

$$R-SH + R'-COCl \longrightarrow R-S-CO-R' + HCl$$

Oxidation of thiols gives products quite different from those of alcohol oxidation. Mild oxidants produce disulphides, e.g.

$$CH_3-SH + I_2 \longrightarrow CH_3-S-S-CH_3 + 2HI$$

and thiols can be quantitatively titrated with iodine. More vigorous oxidation converts the disulphides into sulphonic acids:

$$C_2H_5SH \xrightarrow{HNO_3} [C_2H_5-S-S-C_2H_5] \xrightarrow{HNO_3} C_2H_5SO_3H$$

Thioethers can be alkylated easily to form 'sulphonium' salts (cf. quaternary ammonium salts, p. 123):

$$C_2H_5S-C_2H_5 + C_2H_5-I \longrightarrow \underset{\substack{C_2H_5 \\ \text{Triethylsulphonium iodide}}}{\overset{C_2H_5}{\underset{}{\overset{\oplus}{S}}}}\!\!\!\!\!\!C_2H_5 \quad :\!\!\overset{\ominus}{I}\!\!:$$

Oxidation of thioethers, with powerful oxidising agents, gives successively sulphoxides and sulphones:

$$\underset{\substack{\text{Diethyl}\\\text{sulphide}}}{\overset{C_2H_5}{\underset{C_2H_5}{>}}\!\!\ddot{S}\!:} \xrightarrow{H_2O_2} \underset{\substack{\text{Diethyl}\\\text{sulphoxide}}}{\overset{C_2H_5}{\underset{C_2H_5}{>}}\!\!\overset{\oplus}{S}\!-\!\overset{\ominus}{\ddot{\underset{..}{O}}}\!:} \xrightarrow[\text{or}]{H_2O_2} \underset{\substack{\text{Diethyl}\\\text{sulphone}}}{\overset{C_2H_5}{\underset{C_2H_5}{>}}\!\!\overset{\oplus}{\underset{\oplus}{S}}\!\overset{:\overset{\ominus}{\ddot{O}}:}{\underset{:\overset{\ominus}{\ddot{O}}:}{}}} \longleftrightarrow \overset{C_2H_5}{\underset{C_2H_5}{>}}\!\!S\!\overset{O}{\underset{O}{<}}$$

2.5 Summary

1. Alcohols can be divided into three classes – primary (1°), secondary (2°) and tertiary (3°). They can be prepared by the hydration of olefins in the presence of acid:

$$RCH=CHR + H_2SO_4 \longrightarrow RCH_2-\underset{\underset{OSO_3H}{|}}{CHR} \xrightarrow{H_2O} RCH_2-\underset{\underset{OH}{|}}{CHR}$$

This is a particularly useful method for the preparation of tertiary alcohols.

$$CH_3-\underset{\underset{CH_3}{|}}{C}=CH_2 \xrightarrow[\text{2. } H_2O]{\text{1. } H_2SO_4} CH_3-\underset{\underset{OH}{|}}{\overset{\overset{CH_3}{|}}{C}}-CH_3 \qquad \text{2-Methylpropan-2-ol (3°)}$$

$$not \quad CH_3-\underset{\underset{H}{|}}{\overset{\overset{CH_3}{|}}{C}}-CH_2OH \quad (1°)$$

Alternatively, alcohols can be prepared by reduction of more highly oxidised compounds:

$$\underset{\text{Aldehydes}}{R-\overset{\overset{O}{\|}}{C}H} \longrightarrow \underset{1° \text{ alcohols}}{R-CH_2OH}$$

$$\underset{\text{Ketones}}{R-\overset{\overset{O}{\|}}{C}-R} \longrightarrow \underset{2° \text{ alcohols}}{R-\underset{\underset{OH}{|}}{C}H-R}$$

Finally, a very versatile synthesis of alcohols is the Grignard synthesis:

$$\underset{R'MgX}{\overset{\delta- \;\delta+}{}} + \begin{cases} H_2C=O \quad \text{Methanal} \\ or \\ RCHO \quad \begin{array}{c}\text{Higher}\\\text{aldehydes}\end{array} \\ or \\ RCOR'' \quad \text{Ketones} \end{cases} \xrightarrow{\text{then } \overset{\oplus}{H}} \begin{cases} R'CH_2OH \quad (1°) \\ or \\ R'\underset{\underset{H}{|}}{\overset{\overset{R}{|}}{C}}HOH \quad (2°) \\ or \\ R'\underset{\underset{OH}{|}}{\overset{\overset{R}{|}}{C}}-R'' \quad (3°) \end{cases}$$

2. Alcohols are weakly acidic and form alkoxide anions (salts) when reacted with the metals of groups I and II:

$$ROH \xrightarrow{Na} RO^{\ominus} \, Na^{\oplus} \xrightarrow{R'X} ROR' + NaX$$
$$\text{Metal alkoxide}$$

These alkoxides are nucleophilic and react with alkyl halides (R'X) in the Williamson synthesis of ethers.

Alcohols react readily with carboxylic acids (in the presence of mineral acids), or their derivatives, to form esters:

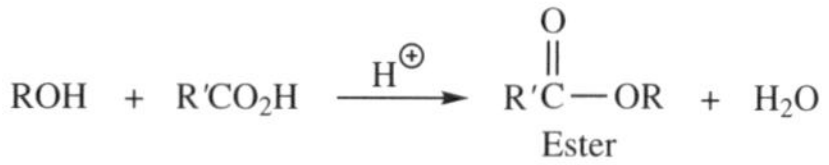

$$ROH + R'CO_2H \xrightarrow{H^{\oplus}} R'\overset{O}{\underset{\|}{C}}-OR + H_2O$$
$$\text{Ester}$$

(R'COCl or R'CO—O—COR' more reactive than R'CO$_2$H)

3. Alcohols are readily oxidised. Primary alcohols (1°) are oxidised to either aldehydes or carboxylic acids, depending upon the conditions used:

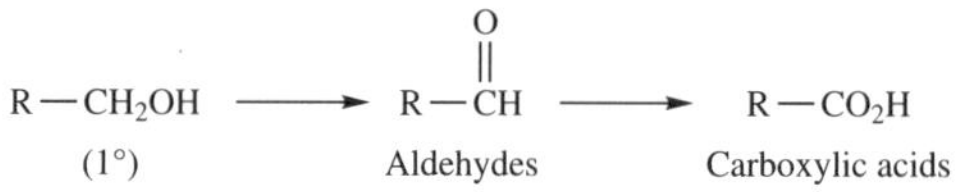

$$R-CH_2OH \longrightarrow R-\overset{O}{\underset{\|}{CH}} \longrightarrow R-CO_2H$$
$$(1°) \qquad \text{Aldehydes} \qquad \text{Carboxylic acids}$$

Secondary (2°) alcohols can be oxidised to ketones, which are relatively resistant to further oxidation, as are tertiary alcohols:

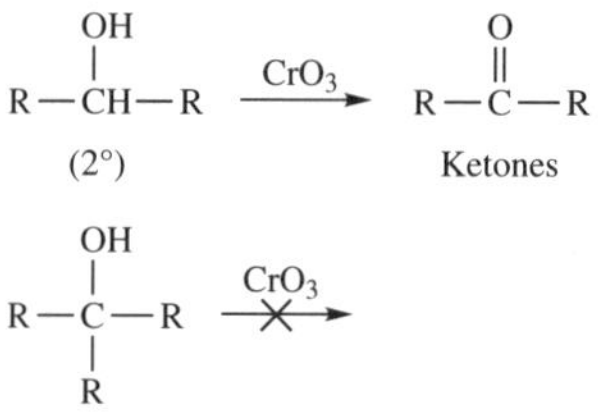

$$R-\underset{(2°)}{\overset{OH}{\underset{|}{CH}}}-R \xrightarrow{CrO_3} R-\overset{O}{\underset{\|}{C}}-R$$
$$\text{Ketones}$$

$$R-\overset{OH}{\underset{\underset{R}{|}}{\underset{|}{C}}}-R \xrightarrow{CrO_3} \times$$

The oxidation of 1,2-diols (polyols) with periodic acid, HIO$_4$, or lead tetra-acetate, Pb(OCOCH$_3$)$_4$, results in cleavage of the C—C bond between the hydroxyl-substituted carbon atoms:

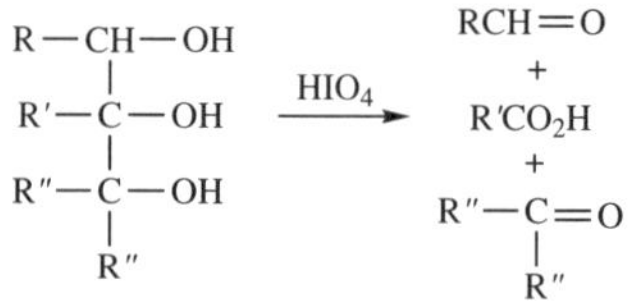

$$\begin{array}{c} R-CH-OH \\ | \\ R'-C-OH \\ | \\ R''-C-OH \\ | \\ R'' \end{array} \xrightarrow{HIO_4} \begin{array}{c} RCH=O \\ + \\ R'CO_2H \\ + \\ R''-C=O \\ | \\ R'' \end{array}$$

4. Alcohols can be converted to alkyl chlorides by reaction with hydrogen halides or non-metal halides:

$$R-OH \xrightarrow[SOCl_2]{HCl \atop or} R-Cl + \begin{cases} H_2O \\ or \\ SO_2 + HCl \end{cases}$$

5. Phenols are prepared from the corresponding diazonium salt and undergo some of the reactions of alcohols: salt formation (occurs more readily since phenols are more acidic than alcohols), ester formation and ether formation.

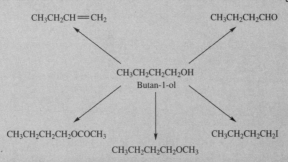

Phenols undergo facile electrophilic substitution at the *ortho-* and *para-* positions.

6. The preparation and reactions of organic sulphur compounds generally resemble those of the oxygen analogues. Notable exceptions are the dialkyl disulphides (formed by oxidation of thiols, and reduced back to the thiols) and thioethers which are readily alkylated and oxidised.

Problems

2.1 Give reagents for the conversion of butan-1-ol into all of the following compounds:

$CH_3CH_2CH=CH_2$ $CH_3CH_2CH_2CHO$

$CH_3CH_2CH_2CH_2OH$
Butan-1-ol

$CH_3CH_2CH_2CH_2OCOCH_3$ $CH_3CH_2CH_2CH_2I$

$CH_3CH_2CH_2CH_2OCH_3$

2.2 What products do you expect to be formed by the reaction of methanol with concentrated sulphuric acid? Give the mechanisms for the reactions involved.

2.3 Treatment of $ClCH_2CH_2OH$ with solid potassium hydroxide gives 'ethylene oxide'. What is the mechanism of this reaction? What would be the product of reacting ethylene oxide with hydriodic acid, HI?

$$CH_2 - CH_2$$
$$\diagdown \diagup$$
$$O$$
Ethylene oxide

2.4 Explain why increasing the number of nitro groups on a phenol leads to an increase in acidity. (Hint: consider the resonance structures possible.)

Compound	pK_a (p. 105)
phenol	9.89
2-nitrophenol	7.12
4-nitrophenol	7.19
2,4-dinitrophenol	4.00
2,4,6-trinitrophenol	0.80

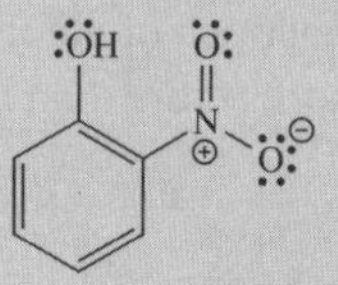

2-Nitrophenol

2.5 Treatment of glycerol (propane-1,2,3-triol) with sodium, followed by methyl iodide (CH$_3$I) gave a pair of isomeric compounds, **A** and **B**. These isomers were separated and reacted with periodic acid, HIO$_4$. Isomer **A** gave methanal (CH$_2$=O) plus **C**, and isomer **B** gave no reaction. Identify **A**, **B** and **C**. What products would be formed by the reaction of **A** and **B** with acetyl chloride (CH$_3$COCl)?

2.6 There are two stable isomers of propanediol (C$_3$H$_8$O$_2$). Draw both isomers and describe how they could be distinguished by a chemical test.

2.7 Give reagents for each of the following conversions:

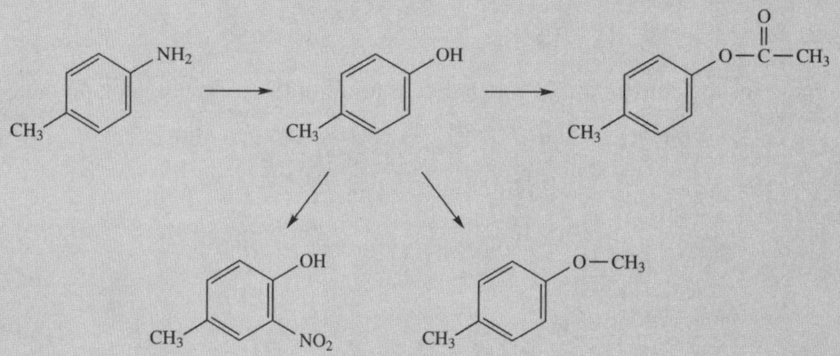

2.8 Compound **D** is converted quantitatively into the alkaloid, salsolin, **E**, in the presence of trace amounts of acid. This reaction is an example of the electrophilic substitution of a phenol but what is the electrophile involved, and how is it formed?

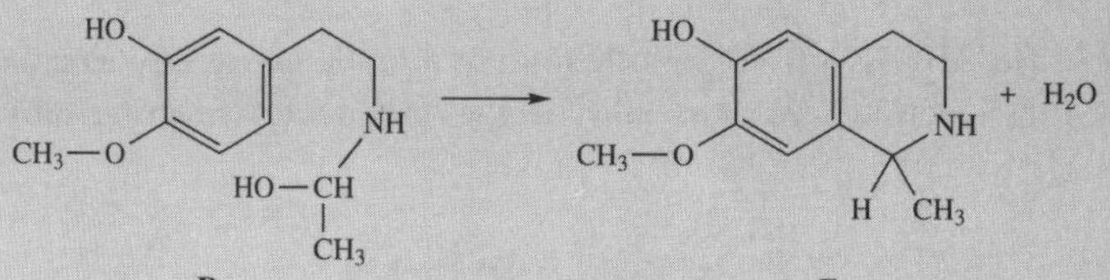

3 Carbonyl compounds: aldehydes and ketones

Topics

3.1 Aldehydes and ketones
3.2 Quinones
3.3 Summary

The carbonyl group, C=O, is the common structural feature of a large number of functional groups, many of which are of great importance in naturally occurring substances, and some of which are illustrated and named below:

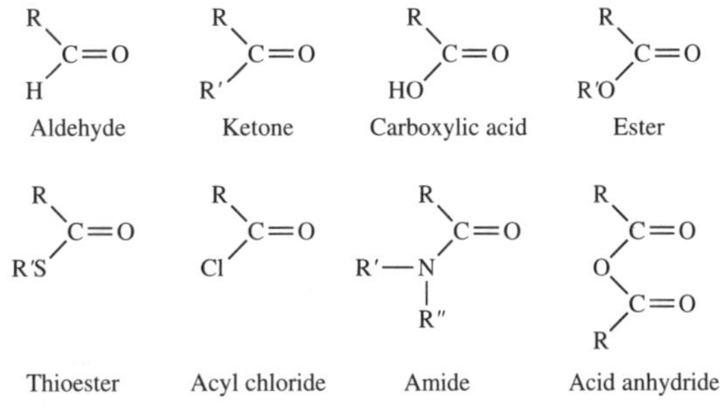

$$\underset{\text{Aldehyde}}{\underset{H}{\overset{R}{>}}\!\!C\!=\!O} \quad \underset{\text{Ketone}}{\underset{R'}{\overset{R}{>}}\!\!C\!=\!O} \quad \underset{\text{Carboxylic acid}}{\underset{HO}{\overset{R}{>}}\!\!C\!=\!O} \quad \underset{\text{Ester}}{\underset{R'O}{\overset{R}{>}}\!\!C\!=\!O}$$

$$\underset{\text{Thioester}}{\underset{R'S}{\overset{R}{>}}\!\!C\!=\!O} \quad \underset{\text{Acyl chloride}}{\underset{Cl}{\overset{R}{>}}\!\!C\!=\!O} \quad \underset{\text{Amide}}{\underset{\underset{R''}{|}}{\underset{R'-N}{\overset{R}{>}}}\!\!C\!=\!O} \quad \underset{\text{Acid anhydride}}{\overset{R}{>}\!C\!=\!O \atop O \atop R\!>\!C\!=\!O}$$

We will consider the chemistry of several classes of carbonyl compounds, but, before doing so, some of the properties of the carbonyl group will be described, since it is the characteristics of this group that are responsible for much of the chemistry of carbonyl-containing functional groups.

The electronic structure of the carbonyl group follows from what has been described previously (p. 9). A carbonyl carbon is joined to three other groups by σ bonds. These employ the three sp² hybrid orbitals of the carbonyl carbon and lie in a plane, 120° apart. The remaining, unused, p orbital overlaps with a p orbital on the oxygen to form a π bond. (The σ bond is formed by the overlap of the C sp² orbital with an O p orbital). Since the sp² orbitals all lie in the same plane the part of the molecule immediately surrounding a carbonyl carbon is flat; the carbon and oxygen of the carbonyl group and the two other atoms attached to the carbon all lie in the same plane. The electron clouds of the π bond lie above and below this plane.

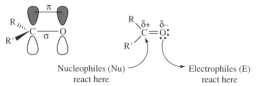

The electronic structure of the carbonyl group

Since the oxygen uses only two p orbitals for bond formation to carbon this leaves two orbitals on the oxygen (an s and a p) which contain the two lone-pairs. Because of the greater electronegativity of oxygen compared with carbon, both the σ and the π bonds of the C=O are greatly polarised (uneven sharing of the bonding electrons), with the *carbon bearing a partial positive charge* (δ+). This positive charge in turn leads to an inductive displacement of electrons along the bonds joining the carbon atom to adjacent groups. These electronic factors are sufficient to explain many of the characteristic reactions of carbonyl compounds. *The lone-pairs of oxygen are the site of electrophilic attack; the carbon of the carbonyl group is the site of nucleophilic attack* (owing to its partial positive charge); and the inductive effect along the bonds between the carbon and neighbouring groups explain some of the characteristic properties of groups adjacent to carbonyl functions.

3.1 Aldehydes and ketones

3.1.1 Nomenclature

Aldehydes (R—CHO)

The IUPAC system of nomenclature indicates the presence of an aldehyde by the suffix '**al**' to the name based upon the parent hydrocarbon. The longest chain carrying the —CHO is said to be the parent structure and is named by replacing the **-e** of the corresponding alkane by **-al**. Substituents are given the lowest possible number but the carbonyl carbon is always C-1.

$$CH_3-CHO \qquad \underset{\beta}{CH_3}-\underset{\alpha\ |}{\overset{2}{CH}}-\overset{1}{CHO}$$
$$Ph$$

Ethanal (acetaldehyde)
Produced in living systems as an intermediate in the enzymatic oxidation of ethanol, or during the alcoholic fermentation of glucose

2-Phenylpropanal (2-phenylpropionaldehyde or α-phenylpropionaldehyde)

An alternative system of nomenclature (which is used to give the common names of aldehydes) replaces the **-ic acid** of the corresponding carboxylic acid by **-aldehyde**. This system is still used in naming aromatic aldehydes.

Substituent positions are indicated by Greek letters (α-, β-, γ-, etc.), with the α-carbon being that which carries the —CHO group (equivalent to the 2-position in the IUPAC system).

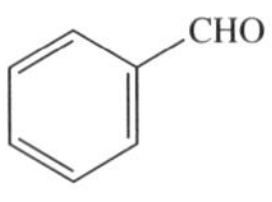

Benzaldehyde 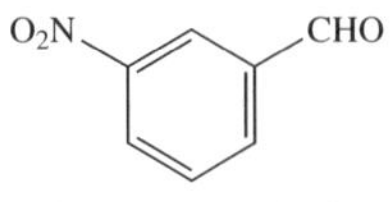   3-Nitrobenzaldehyde 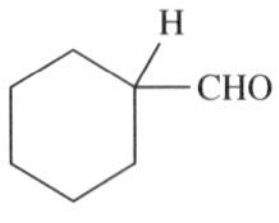   Cyclohexanecarboxaldehyde

Ketones (R—CO—R′)

Ketones are described by the prefix '**oxo**' or the suffix '**one**'. In addition, simple ketones are frequently given trivial (common) names indicating the two groups attached to the carbonyl group.

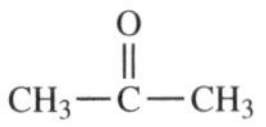

Propanone (acetone) Pentan-2-one   Cyclohexanone
occurs in the urine of people (methyl propyl ketone) (oxocyclohexane)
with diabetes

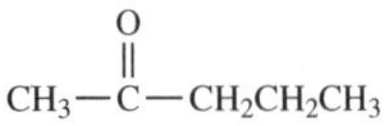

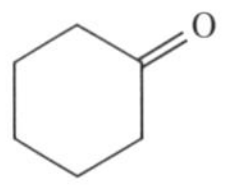

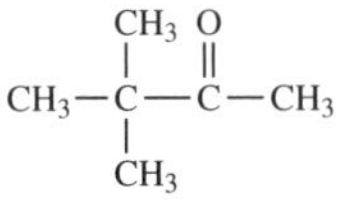

3,3-Dimethylbutan-2-one   Acetophenone   Benzophenone
(*tert*-butyl methyl ketone)

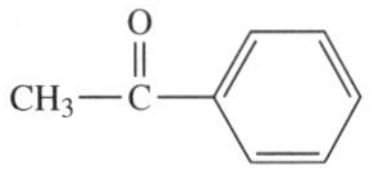

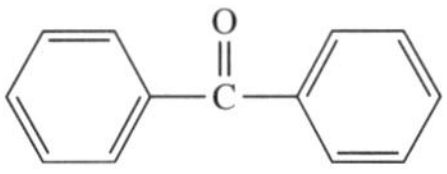

Once again, common names are used for the aromatic ketones and ketones in which C=O is attached to a benzene ring are known as **-phenones**.

3.1.2 Preparation

Aldehydes

Oxidation of primary alcohols. The classical method of preparation of aldehydes is the oxidation of a primary alcohol. However, care must be taken that the *aldehyde produced is not oxidised further to a carboxylic acid*. Chromic acid (acidified sodium dichromate, Jones' reagent) is frequently used in the laboratory, but industrially oxygen and catalysts are employed. Chromic acid is a strong oxidising agent and is only suitable for the preparation of aldehydes that have a lower boiling point than the primary alcohol and can be removed from the reaction mixture by distillation.

$$R-CH_2OH \xrightarrow[H^\oplus]{Na_2Cr_2O_7} R-CHO \longrightarrow RCO_2H$$

A *milder oxidising agent* is pyridinium chlorochromate (PCC) which is frequently the reagent of choice for the preparation of aldehydes:

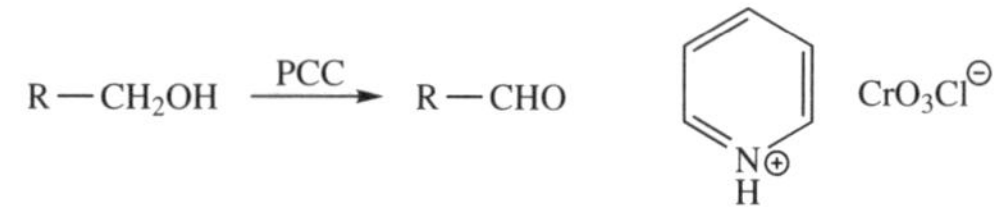

82 Carbonyl compounds: aldehydes and ketones

$$R-CH_2OH \xrightarrow{PCC} R-CHO$$

Pyridinium chlorochromate (PCC)

Citronellol $\xrightarrow{PCC}$ Citronellal

The oxidation of alcohols in biological systems uses complex coenzymes such as nicotinamide-adenine dinucleotide (NAD+) and nicotinamide-adenine dinucleotide phosphate (NADP+) as the oxidising agents. The chemically active group in both these cases is the nicotinamide moiety, which can undergo a reversible reduction in which the pyridinium ring is reduced to a dihydropyridine by reaction with a hydride ion (or the chemically equivalent hydrogen ion and two electrons). These coenzymes and their reduced derivatives can therefore act as acceptors or donors of H⁻ or electrons.

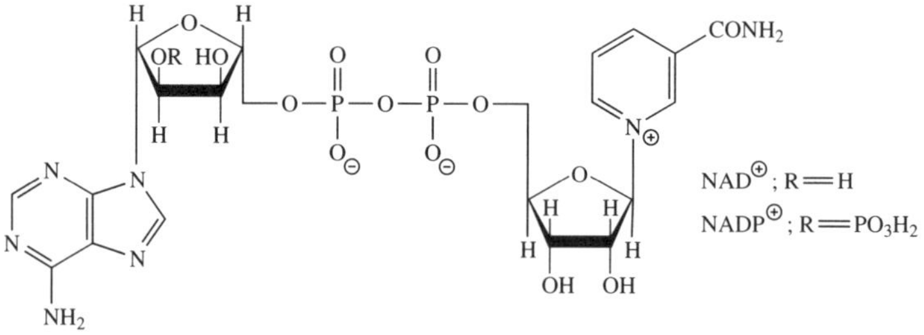

NAD⊕; R=H
NADP⊕; R=PO₃H₂

(NAD = adenine-D-ribose-phosphate-pho sphate-D-ribose-nicotinamide)

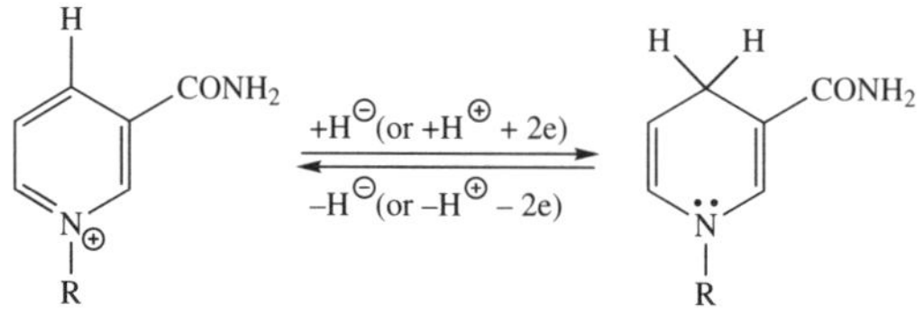

Pyridinium cation $\xrightleftharpoons[-H^{\ominus} (or -H^{\oplus} - 2e)]{+H^{\ominus} (or +H^{\oplus} + 2e)}$ 1,4-Dihydropyridine derivative

The shorthand representation of these reactions is:

$$H^+ + 2e + NAD^+ \rightleftharpoons NADH$$

$$H^+ + 2e + NADP^+ \rightleftharpoons NADPH$$

Although these two coenzymes differ only marginally in their structure, there is absolute specificity for one or other in enzymic reactions. In general, degradative processes involve NAD+ and NADH, while synthetic processes utilise NADP+ and NADPH.

Hydrolysis of gem-*dihaloalkanes.* This method is particularly useful for the preparation of aromatic aldehydes since the *gem*-dihaloalkanes are readily available from the chlorination of methylbenzenes:

$$\text{Ar}-\text{CH}_3 \xrightarrow[\text{heat}]{\text{Cl}_2,\ \text{light}} \text{Ar}-\text{CHCl}_2 \xrightarrow{\text{H}_2\text{O}} \text{Ar}-\text{CHO}$$

The intermediate in the hydrolysis of the dichloroalkane is the *gem*-diol which readily loses water to give the aldehyde:

$$\text{Ar}-\text{CH(OH)}_2 \longrightarrow \text{Ar}-\text{CHO}$$

Oxidative cleavage of alkenes

1. **Ozonolysis**. Alkenes react with ozone to form ozonides, which on reduction give carbonyl compounds:

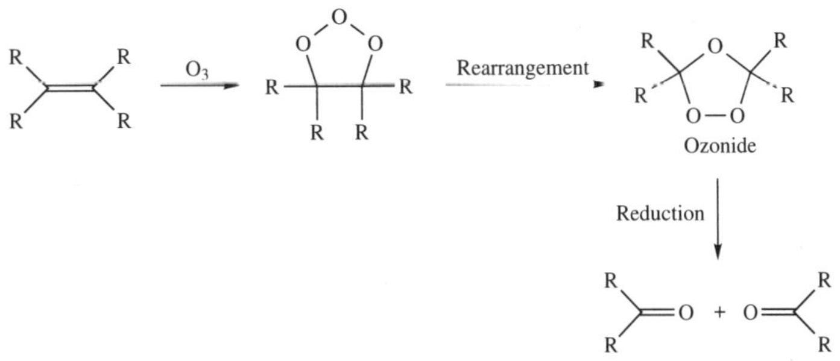

Thus, if one (or more) of the R groups attached to the alkene is hydrogen (vinylic hydrogen), (an) aldehyde(s) will be obtained:

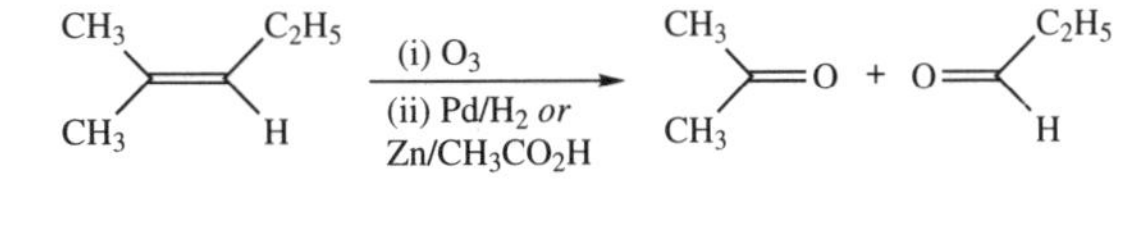

(Note that the carbon–carbon double bond has been broken and in its place are two carbon–oxygen double bonds.)

2. **Oxidation of *vic*-diols.** This method is closely related to ozonolysis since the *vic*-diols are obtained by the oxidation of an alkene with potassium permanganate in basic solution. Oxidation of the diol by periodic acid, HIO_4, or lead tetraacetate, $Pb(OCOCH_3)_4$, then gives the carbonyl compounds. This two-step process is thus directly analogous to ozonolysis followed by reduction of the ozonide.

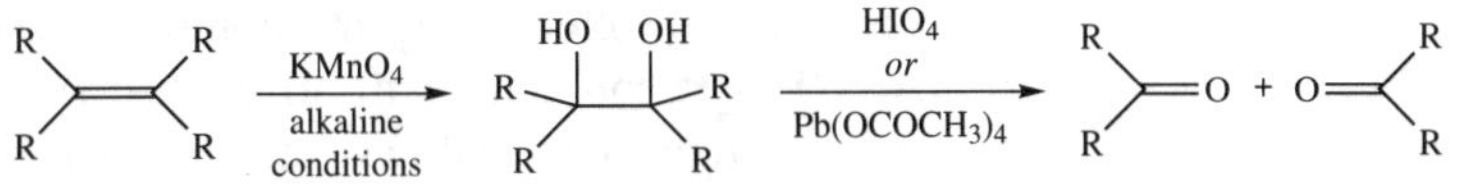

Ketones
Oxidation of secondary alcohols. This process is directly analogous to the formation of aldehydes from primary alcohols, except that in this case no precautions need be taken against further oxidation (of the ketone) so that acidified sodium dichromate is the reagent of choice:

[Reaction scheme: (−)-Menthol → (−)-Menthone using Na₂Cr₂O₇ / H⁺]

As mentioned previously, the oxidation of alcohols in biological systems uses the enzymes NAD⁺ and NADP⁺.

Friedel–Crafts acylation. This is an example of an electrophilic aromatic substitution and is a good method for the preparation of aromatic ketones. Friedel–Crafts acylation involves the reaction of a (substituted) benzene with an acid chloride, in the presence of a Lewis acid catalyst (usually $AlCl_3$). The electrophile in this case can be thought to be the acylium ion, $R—C\equiv O^+$.

$$Ar—H + R—COCl \xrightarrow{AlCl_3} Ar—CO—R + HCl$$

Oxidative cleavage of alkenes.

Hydration of alkynes. The hydration (addition of water) of an alkyne, which is catalysed by mercury salts, gives ketones (except for the hydration of ethyne which produces an aldehyde).

$$R—C\equiv C—H \xrightarrow[H^+/Hg^{2+}]{H_2O} R—C(OH)=CH_2 \rightleftharpoons R—CO—CH_3$$

Alkyne — Enol tautomer — Keto tautomer

3.1.3 Reactions

Oxidation

As mentioned previously, aldehydes are readily oxidised to the corresponding carboxylic acid while ketones are not. As well as the usual powerful oxidising agents such as acidified sodium dichromate, potassium permanganate and nitric acid, the oxidation of aldehydes can be performed by a number of relatively weak oxidants. Fehling's solution (an alkaline solution of a copper(II) tartrate complex ion), Benedict's solution (alkaline copper(II) citrate complex), or Tollen's reagent (ammoniacal silver nitrate, i.e. weakly alkaline $[Ag(NH_3)_2]^+$ solution), will all oxidise simple aldehydes, being reduced to copper(I) oxide or metallic silver (the '**silver mirror**' test). These reagents do not oxidise simple ketones (cf. carbohydrates, p. 145). (NB **Strongly** alkaline $[Ag(NH_3)_2]^+$ solution will also oxidise many simple ketones, with the deposition of metallic silver.)

$$R—CHO \xrightarrow{[O]} R—COOH$$

$$CH_3CHO + 2Cu^{2\oplus} + 3OH^{\ominus} \longrightarrow CH_3CO_2^{\ominus} + 2Cu^{\oplus} + 2H_2O$$

$$2Cu^{\oplus} + 2OH^{\ominus} \longrightarrow Cu_2O\downarrow + H_2O$$

The oxidation of simple ketones is a much more difficult reaction; prolonged, vigorous oxidation will, however, degrade a ketone by cleavage of the C—CO bond to give a mixture of carboxylic acids.

$$CH_3CH_2CH_2-C(=O)-CH_2CH_3 \xrightarrow[H^{\oplus}]{Na_2Cr_2O_7} \begin{cases} CH_3CH_2CH_2CO_2H + HO_2CCH_3 \\ \\ CH_3CH_2CO_2H + HO_2CCH_2CH_3 \end{cases}$$

Reduction

Reduction of aldehydes and ketones can lead to either the parent alcohol or hydrocarbon, depending upon which method of reduction is employed. Reduction with a complex metal hydride, such as lithium aluminium hydride (LiAlH$_4$) or sodium borohydride (NaBH$_4$), is an example of an irreversible nucleophilic addition to the carbonyl group, the nucleophile being the hydride (H$^-$) ion:

(Aldehydes give primary alcohols; ketones give secondary alcohols.)

Metal–acid reduction (e.g. Zn/HCl, Sn/HCl) and catalytic hydrogenation (e.g. Ni/H$_2$) also usually give the alcohol, but the use of zinc amalgam and hydrochloric acid (Clemmensen reduction), or hydrazine then a strong base (Wolff–Kishner reduction), gives the parent hydrocarbon:

Clemmensen reduction (for compounds sensitive to base)

Wolff–Kishner reduction (for compounds sensitive to acid)

The reaction of aldehydes and ketones with nucleophiles

Aldehydes and ketones form a series of characteristic derivatives by 'addition reactions' (i.e. reactions in which the molecular formula of the product is the sum of the molecular formulae of the reagents) and 'condensation reactions' (i.e. reactions in which combination of the reagents occurs with the elimination of water or an alcohol). Both of these types of reaction have a common initial pathway commencing with nucleophilic attack on the carbonyl carbon. *(Nucleophilic acyl substitution – see p. 111 – does not take place for aldehydes and ketones since the leaving group would be H or R, both of which are poor leaving groups.)*

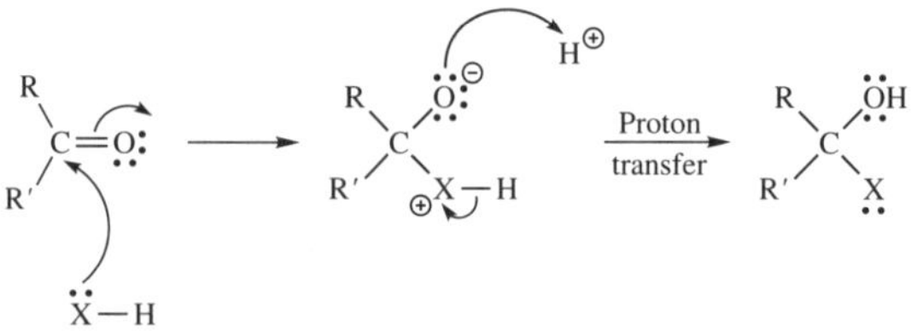

86 Carbonyl compounds: aldehydes and ketones

Overall reaction: R—CO—R' + HX ⟶ RC(OH)X—R'
The mechanism of addition of HX to RCOR'

The polarisation of the carbonyl group leaves the carbon atom with a partial positive charge. The nucleophilic group X of the reagent H—X donates a lone-pair of electrons to the carbon atom and, since carbon is tetravalent, the weak π bond breaks with the electron pair moving onto the oxygen atom, resulting in a doubly charged intermediate. Proton transfer to and from other intermediates (or the solvent) then takes place. (Alternatively, if H—X ionises to H^+ and X^- the initial nucleophilic attack may be by X^- rather than HX, followed by addition of H^+ to the negatively charged oxygen atom.)

The fate of the product, RC(OH)X—R', depends upon the nature of the nucleophile, X—H. The formation of the product may be reversible, irreversible, or it may lead to the elimination of water (see Table 3.1).

Table 3.1
Addition reactions of aldehydes and ketones (R—CO—R')

Nucleophile, HX	Product of reaction	Comments
H—CN (+ KCN cat.)	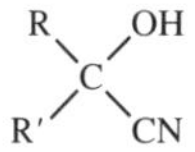 RR'C(OH)CN Cyanohydrin	Nucleophilic attack by CN^-; reversible*
H—NH$_2$	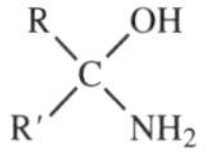 RR'C(OH)NH$_2$	Not usually isolable; reversible
H—OH	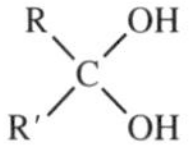 RR'C(OH)OH Gem-diol	Not usually isolable; reversible
H—OR"	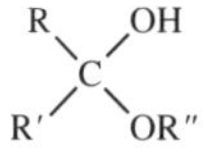 RR'C(OH)OR" Hemiacetal	Not usually isolable; reversible
H—OSO$_2^\ominus$Na$^\oplus$	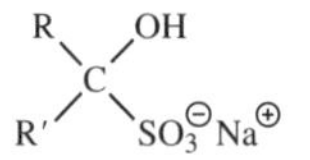 RR'C(OH)SO$_3^\ominus$Na$^\oplus$ Sodium bisulphite adduct	Reversible

*Under the reaction conditions.

Cyanohydrin (hydroxynitrile) formation. This is an important reaction in sugar (carbohydrate) chemistry and its reversibility has been employed in nature by certain plants and insects as a defence mechanism. For example, the stones of apricots and peaches contain a group of compounds called the cyanogenic glycosides, which consist of benzaldehyde cyanohydrin with simple sugars attached. When eaten, the sugar unit is cleaved off, and poisonous HCN is released.

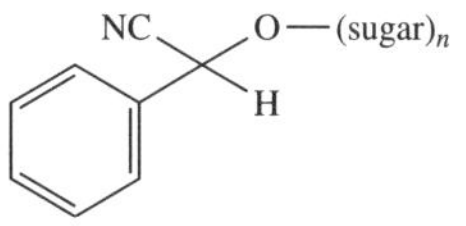

In addition, since the acid-catalysed hydrolysis of nitriles gives carboxylic acids (see p. 104), the cyanohydrins are useful synthetic intermediates.

The addition of ammonia and its derivatives. Condensation reactions occur between aldehydes or ketones and many compounds containing the —NH$_2$ group (derivatives of ammonia):

The mechanism for this reaction involves the initial addition of H$_2$N—Y (see mechanism shown above). Under weakly acidic conditions (e.g. an acetate buffer solution), protonation of the oxygen atom of the hydroxyl group is followed by the elimination of water and loss of a proton from the NH group:

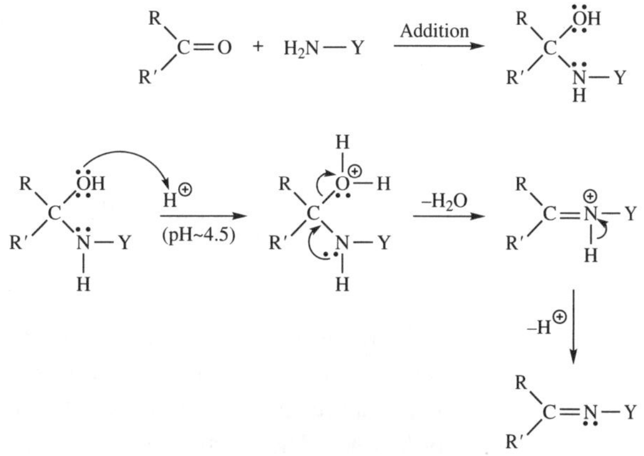

It may seem surprising that in the step following the formation of the addition compound, protonation occurs on the oxygen atom rather than on nitrogen, when amines are known to be more basic than alcohols. Protonation of nitrogen undoubtedly can occur, but the only reaction possibilities open to the resultant compound are loss of the newly acquired proton, or concerted elimination of a proton and H$_2$N—Y to re-form the original reagents. Although protonation of oxygen is not very favoured, it leads irreversibly (under the reaction conditions) to the condensation compound (Table 3.2).

88 Carbonyl compounds: aldehydes and ketones

Table 3.2
Typical condensation reactions of aldehydes and ketones (R—CO—R′)

Ammonia derivative, H$_2$N—Y	Product
H$_2$N—R″ (amine)	R\R′C=N—R″ (imine; Schiff's base)
H$_2$N—OH (hydroxylamine)	R\R′C=N—OH (oxime)
H$_2$N—NH$_2$ (hydrazine)	R\R′C=N—NH$_2$ (hydrazone)
H$_2$N—NHAr (arylhydrazine)	R\R′C=N—NHAr (arylhydrazone)
H$_2$N—NHCO.NH$_2$ (semicarbazide)	R\R′C=N—NHCO.NH$_2$ (semicarbazone)

Amines, hydroxylamine and hydrazine are reasonably strong nucleophiles and, as such, do not require acid catalysis of their initial addition to the carbonyl group. Weaker nucleophiles such as arylhydrazines (e.g. Ar = 2,4-dinitrophenyl) require acid catalysis to activate the carbonyl group towards nucleophilic attack:

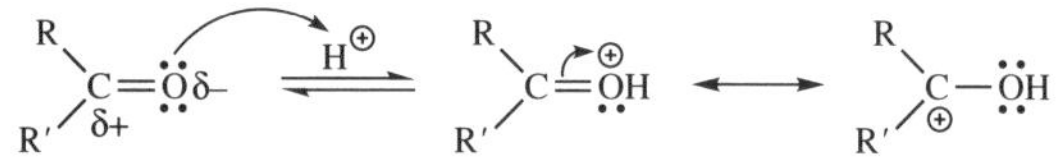

$$\underset{R'}{\overset{R}{>}}C=\overset{\delta-}{\ddot{O}}\overset{\delta+}{} \;\overset{H^\oplus}{\rightleftharpoons}\; \underset{R'}{\overset{R}{>}}C\overset{\oplus}{=}\overset{..}{\overset{..}{O}}H \;\longleftrightarrow\; \underset{R'}{\overset{R}{>}}\overset{\oplus}{C}—\overset{..}{\overset{..}{O}}H$$

Protonation of the carbonyl oxygen produces a resonance-stabilised carbocation in which one of the resonance forms, the most stable, has a full positive charge on carbon (compared with a partial positive charge in the aldehyde or ketone). The electrophilicity of the carbonyl group has thus been increased and it is more susceptible to attack by the weaker nucleophile.

The products of the above reactions can be hydrolysed by boiling with dilute mineral acid (e.g. 1 M HCl, pH = 0) with the mechanism of the hydrolysis being exactly the reverse of the formation. Imines are particularly susceptible to hydrolysis, unless one of the substituents (R, R′ or R″) is aryl, in which case they are known as Schiff's bases.

Gem-*diol* formation. The hydration (addition of water) to carbonyl compounds can be either acid- or base-catalysed.

1. **Acid catalysis.** Water is a poor nucleophile and thus requires prior protonation of the carbonyl group (see previous section) to produce a resonance-stabilised carbocation, which is more susceptible to nucleophilic attack:

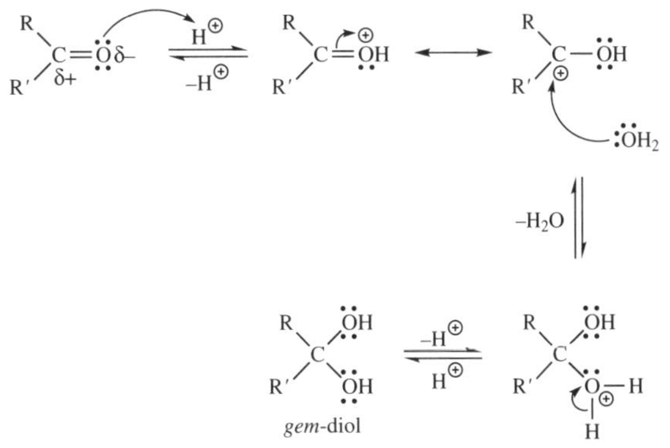

   As can be seen from this mechanism, this process is reversible, with the mechanism being the exact opposite of the hydration.

2. **Base catalysis.** Under basic conditions the carbonyl group is attacked by a good nucleophile (OH⁻) and the adduct is then protonated.

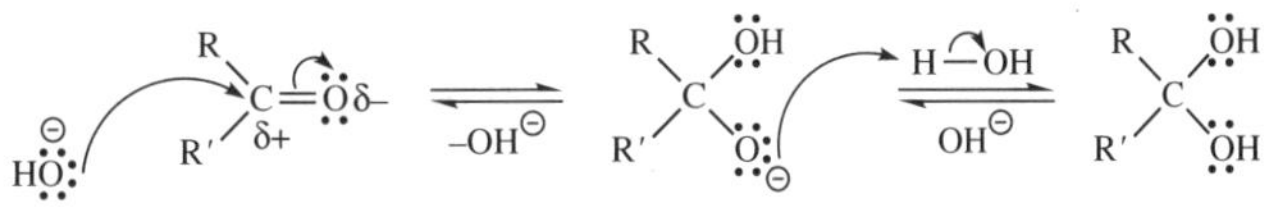

Hemiacetal/acetal formation. Like water, alcohols are poor nucleophiles and so require acid catalysis for hemiacetal/acetal formation. The mechanism for hemiacetal formation (which is an important process in sugar chemistry) is directly analogous to that for *gem*-diol formation:

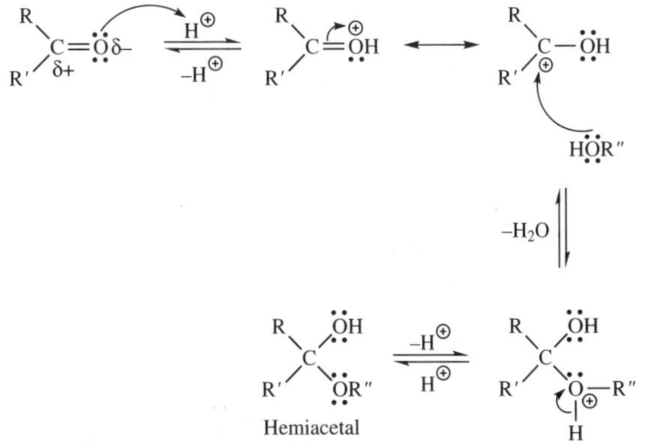

Once again, this is a reversible process; protonation of the hemiacetal can take place on either oxygen (hydroxy or alkoxy). Protonation of the alkoxy group is the first step in the removal of the alkoxy group to give the aldehyde.

Protonation of the hydroxy group, however, converts it into a good leaving group (lost as water) and leads, via a resonance-stabilised carbocation, to the formation of an acetal (derived from an aldehyde):

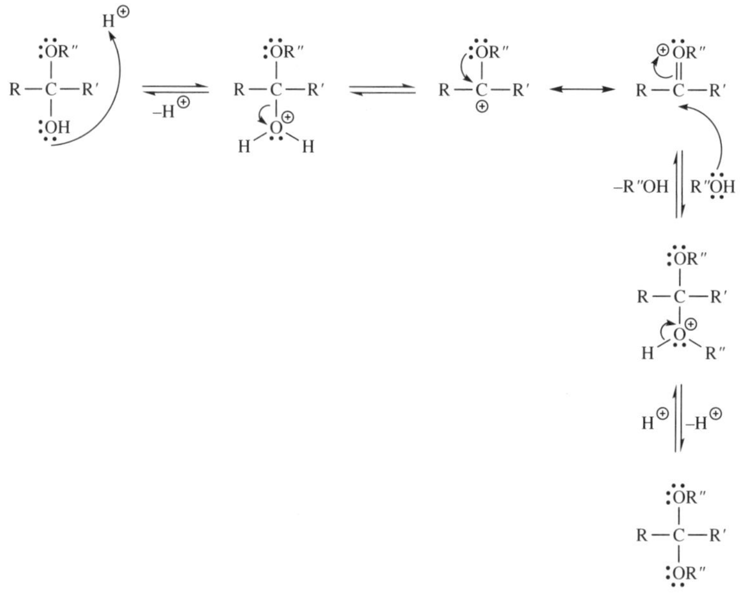

[Reaction scheme showing stepwise mechanism for acetal formation from a hemiacetal via protonation, loss of water, addition of R"OH, and deprotonation to give the Acetal]

Ketals (derived from ketones) are not as readily formed as acetals but cyclic ketals can often be made from 1,2- or 1,3-diols:

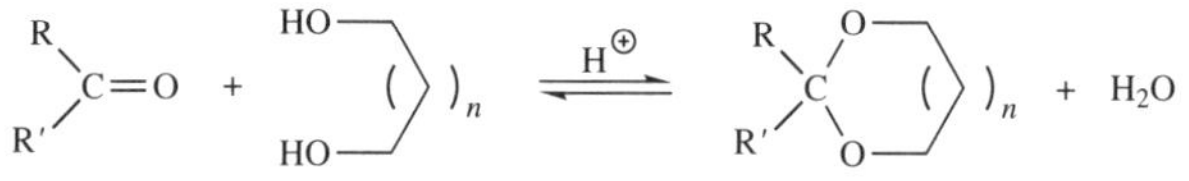

[Reaction scheme: ketone + HO–(CH₂)ₙ–OH (n = 0, 1) with H⁺ catalyst gives cyclic ketal + H₂O]

Hemithioacetals (ketals) and thioacetals (ketals) are more readily formed than their oxo analogues, by the reaction of a thiol (R"SH) with the corresponding aldehyde or ketone (anhydrous zinc chloride is often used as the catalyst).

Addition of sodium bisulphite. Simple carbonyl compounds (aldehydes, methyl ketones and some cyclic ketones) react with bisulphite ion to form an

$$HO^\ominus + HSO_3^\ominus \rightleftharpoons H_2O + SO_3^{2\ominus}$$

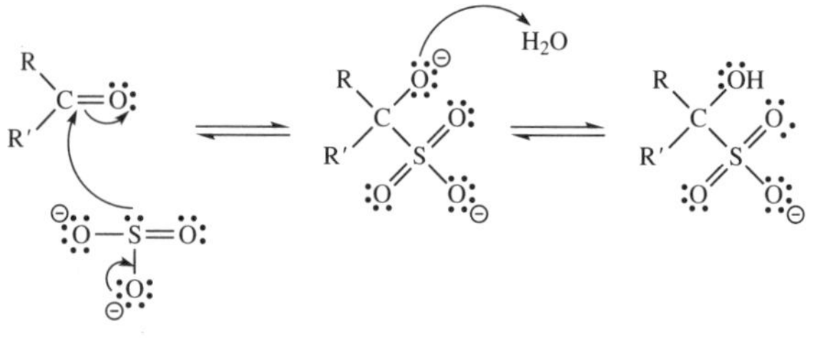

[Mechanism showing addition of sulfite to carbonyl carbon, followed by protonation to give the α-hydroxysulfonate]

adduct. Once again, this is a reversible process. The nucleophile involved is probably the sulphite ion, which is formed in solution from the reaction of hydroxide ion with bisulphite ion.

Keto–enol tautomerism and the aldol reaction

In the presence of bases, compounds containing an α-hydrogen are converted into anions by removal of this proton, which is facilitated by the following:

1. **The electron-withdrawing effect of the carbonyl group.** As mentioned previously, the greater electronegativity of the oxygen atom causes polarisation of both C—O bonds of the carbonyl group, with the carbon atom bearing a partial positive charge. This positive charge produces an inductive displacement of electrons along the C—CO bond which produces a corresponding, weaker, displacement along the C—H bond, making the proton acidic (electron-poor) and so easily removed by a base.

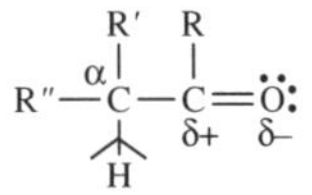

2. **Resonance stabilisation of the anion.** The negative charge on the anion produced is not centred on the α-carbon atom, since the orbital containing the two electrons is conjugated (p. 38) to the π bond of the carbonyl group. The effect of this conjugation is to distribute the negative charge over more than one atom and, in particular, onto the electronegative oxygen atom, and such a resonance-stabilised (mesomeric) anion is always much easier to produce than a corresponding anion in which the negative charge is localised on one atom.

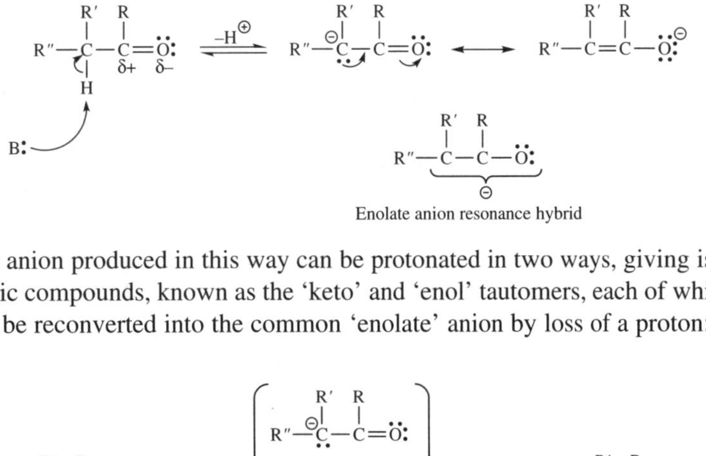

Enolate anion resonance hybrid

The anion produced in this way can be protonated in two ways, giving isomeric compounds, known as the 'keto' and 'enol' tautomers, each of which can be reconverted into the common 'enolate' anion by loss of a proton:

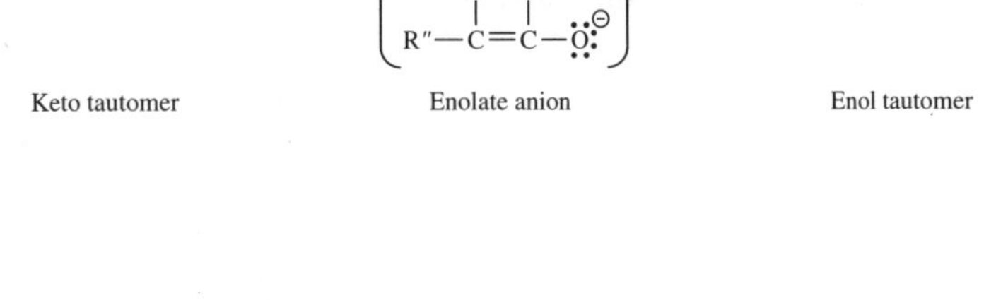

Keto tautomer   Enolate anion   Enol tautomer

Thus, in the presence of base, any aldehyde or ketone with α-hydrogens will be in equilibrium with its enol isomer, conversion occurring via a low equilibrium concentration of the enolate anion. In practice, although strong bases (e.g. $C_2H_5O^-$) are needed to obtain high concentrations of enolate anions, even weak bases, such as the alkaline surface of glass, catalyse the interconversion of keto and enol tautomers, so that any normal sample of an aldehyde or ketone will contain a low concentration of the enol (if such an isomer is possible). With simple aldehydes and ketones the proportion of enol is usually <1 per cent, but may rise to over 50 per cent. For example, if there is a possibility of conjugation and/or internal hydrogen bonding in the enol, the equilibrium will become more balanced.

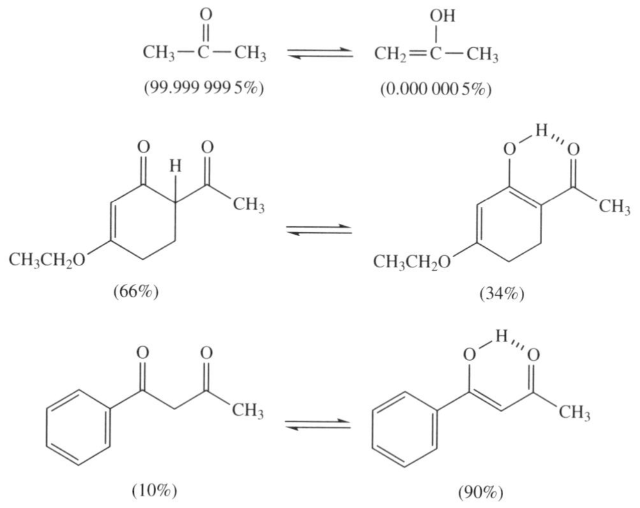

$$CH_3-\underset{\underset{(99.999\,999\,5\%)}{}}{\overset{\overset{O}{\|}}{C}}-CH_3 \rightleftharpoons \underset{\underset{(0.000\,000\,5\%)}{}}{CH_2=\overset{\overset{OH}{|}}{C}-CH_3}$$

In addition to the base-catalysed interconversion of keto and enol isomers, isomerisation is also catalysed by acid. The intermediate for acid-catalysed interconversion is the species produced by protonation of the carbonyl group, which can lose a proton from the oxygen atom to form the keto isomer, or from the α-position to form the enol:

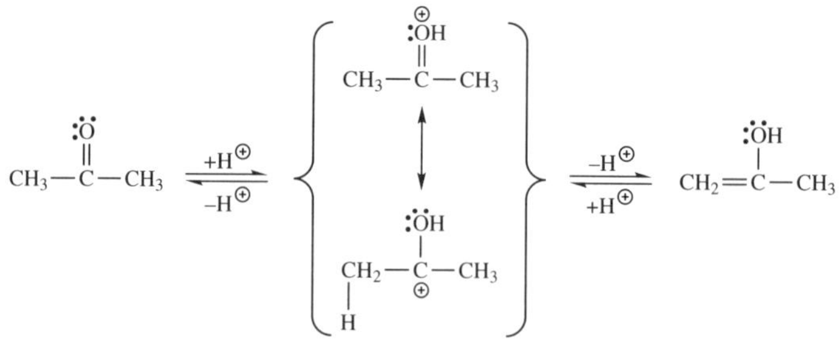

Constitutional isomers which are rapidly interconverted are known as '**tautomers**', and the phenomenon of tautomerism is most commonly encountered in compounds which have a 'mobile' hydrogen atom. It is important to distinguish between tautomerism and resonance. **Tautomers** are different compounds, which are normally rapidly interconverted, whose structures differ markedly in the arrangement of atoms, and are capable of isolation as distinct compounds.

Canonical structures (resonance forms) of mesomeric compounds differ only in their electronic distribution; they are different ways of describing the same species.

The enolate anions of aldehydes and ketones are intermediates in a number of reactions, the most important of which, from a biochemical point of view, is the aldol reaction.

The aldol reaction. In the presence of bases, most aldehydes and ketones are converted into dimers (compounds whose molecular formulae are twice that of the starting material – the monomer):

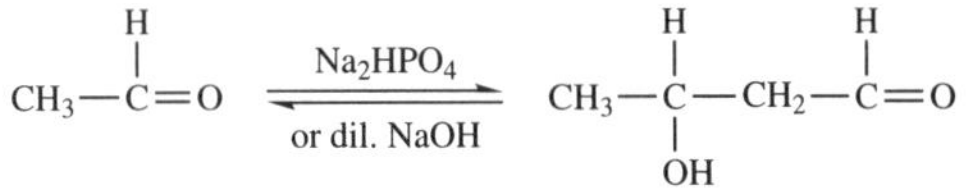

$$CH_3-\overset{\overset{\displaystyle H}{|}}{C}=O \quad \underset{\text{or dil. NaOH}}{\overset{Na_2HPO_4}{\rightleftharpoons}} \quad CH_3-\underset{\underset{\displaystyle OH}{|}}{\overset{\overset{\displaystyle H}{|}}{C}}-CH_2-\overset{\overset{\displaystyle H}{|}}{C}=O$$

Ethanal (acetaldehyde) → 3-Hydroxybutanal (aldol)

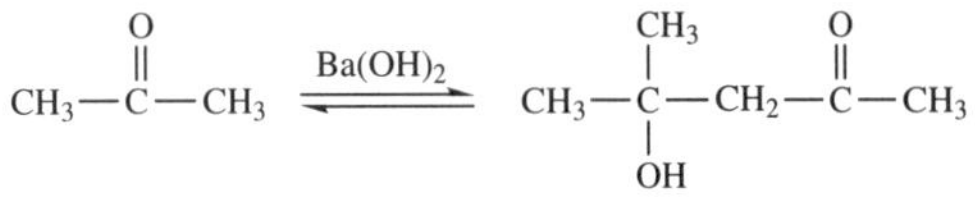

$$CH_3-\overset{\overset{\displaystyle O}{\|}}{C}-CH_3 \quad \overset{Ba(OH)_2}{\rightleftharpoons} \quad CH_3-\underset{\underset{\displaystyle OH}{|}}{\overset{\overset{\displaystyle CH_3}{|}}{C}}-CH_2-\overset{\overset{\displaystyle O}{\|}}{C}-CH_3$$

Acetone → 2-Hydroxy-2-methylpentan-4-one

This reaction, known as the 'aldol addition' after the common name of the ethanal dimer, occurs via the enolate anion, which, acting as a nucleophile, attacks the carbonyl group of another aldehyde or ketone molecule. Protonation of the alkoxide anion (by solvent or by a carbonyl molecule) gives the product, in what is a completely reversible process:

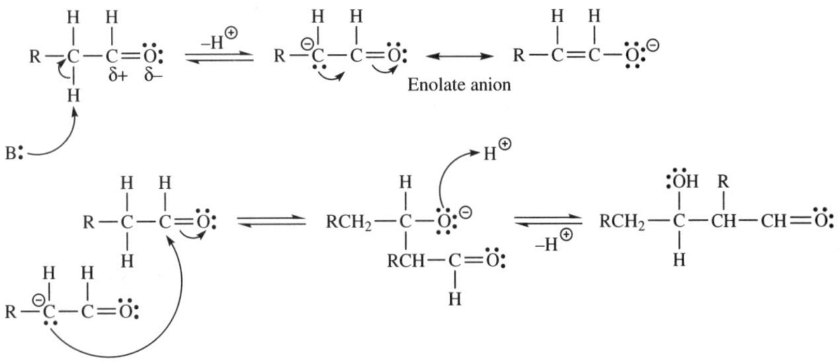

The aldol addition is also possible under acid catalysis, the reactive species in this case being the enol rather than the enolate anion. Under acid catalysis the final product usually undergoes dehydration to give the unsaturated, conjugated product.

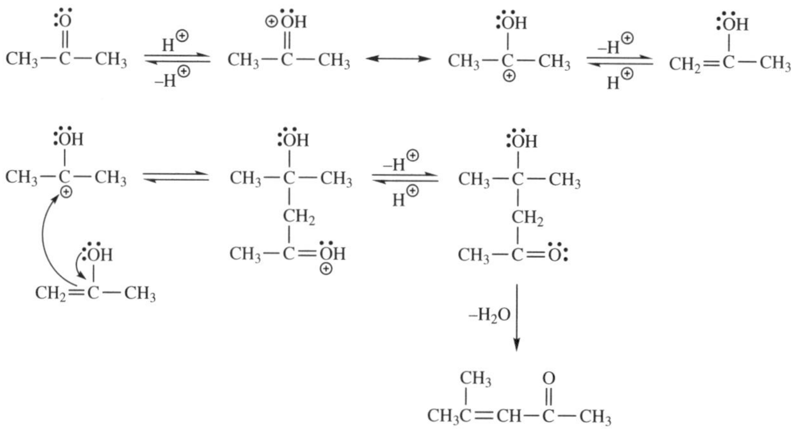

94 Carbonyl compounds: aldehydes and ketones

[Reaction scheme showing acid-catalyzed aldol condensation mechanism of acetone leading to 4-Methylpent-3-en-2-one]

4-Methylpent-3-en-2-one

Examples of the aldol addition are very important reactions in biochemistry, particularly in carbohydrate metabolism. One example is the combination of C_3 and C_4 sugars to give a C_7 sugar, which is a reversible reaction catalysed by the enzyme 'aldolase'.

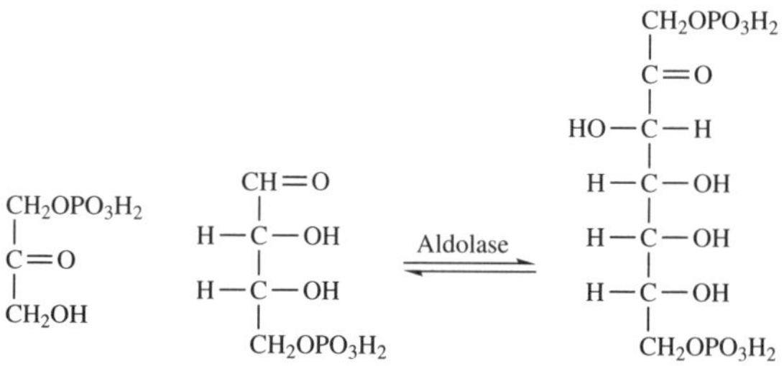

[Scheme: Dihydroxyacetone phosphate + D-Erythrose-4-phosphate ⇌ (Aldolase) D-Sedoheptulose-1,7-diphosphate]

Addition of Grignard reagents

Grignard reagents, which are conveniently prepared from an alkyl halide, RX, and magnesium under anhydrous conditions, are carbon nucleophiles and so react with aldehydes and ketones to form a bond between the carbon that is bound to magnesium and the carbonyl carbon:

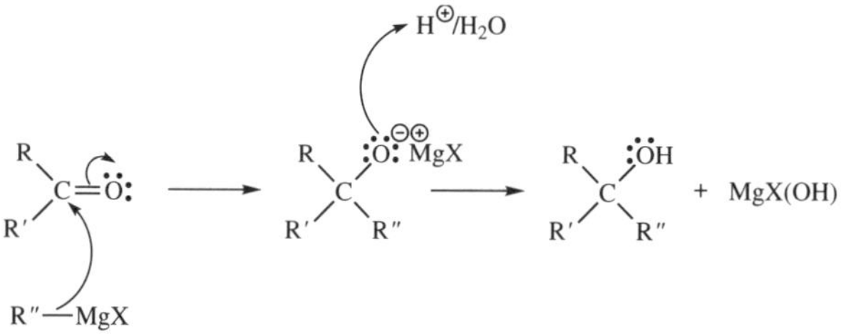

[Mechanism scheme: R,R'C=O + R''—MgX → R,R',R''C—O⁻MgX⁺ → (H⁺/H₂O) R,R',R''C—OH + MgX(OH)]

This is the classical method of preparing alcohols since a variety of alcohols can be prepared depending upon the alkyl halide and the carbonyl compound used.

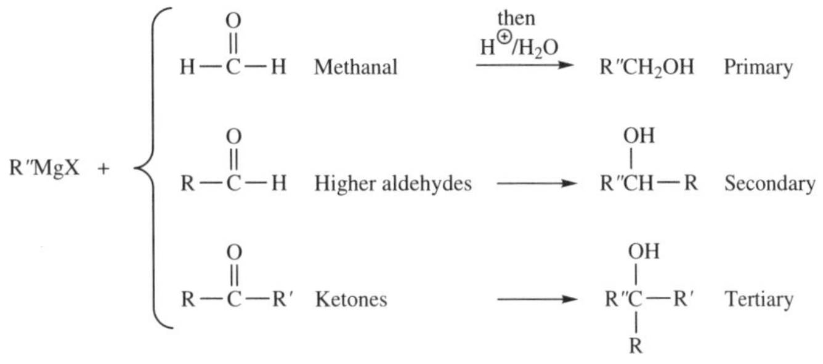

$$R''MgX + \begin{cases} \underset{\text{Methanal}}{H-\overset{\overset{O}{\|}}{C}-H} \\ \underset{\text{Higher aldehydes}}{R-\overset{\overset{O}{\|}}{C}-H} \\ \underset{\text{Ketones}}{R-\overset{\overset{O}{\|}}{C}-R'} \end{cases} \xrightarrow[]{\text{then } H^{\oplus}/H_2O} \begin{array}{l} R''CH_2OH \quad \text{Primary} \\ \\ R''\overset{\overset{OH}{|}}{C}H-R \quad \text{Secondary} \\ \\ R''\overset{\overset{OH}{|}}{\underset{\underset{R}{|}}{C}}-R' \quad \text{Tertiary} \end{array}$$

The Grignard synthesis of alcohols

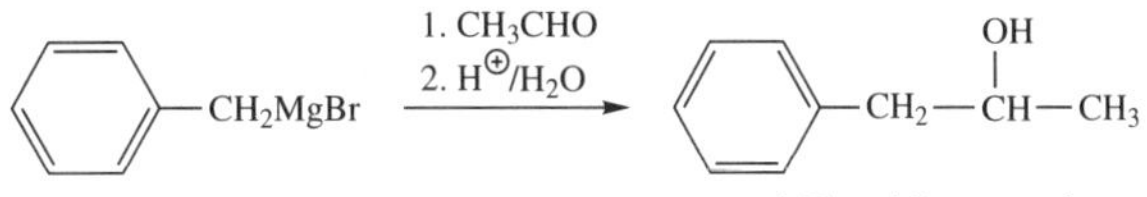

Ph–CH₂MgBr →(1. CH₃CHO; 2. H⊕/H₂O)→ Ph–CH₂–CH(OH)–CH₃

1-Phenyl-2-propanol

This addition, like that of H⁻ (see p. 85), is irreversible under the reaction conditions.

3.2 Quinones

Quinones are the oxidation products of dihydroxybenzenes, from which they can be prepared using weak oxidising agents such as iron(III) chloride or even air:

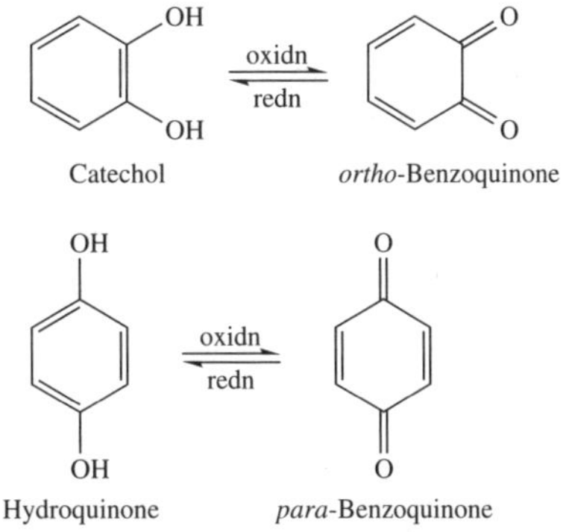

Catechol ⇌ (oxidn/redn) ortho-Benzoquinone

Hydroquinone ⇌ (oxidn/redn) para-Benzoquinone

Quinones exhibit very few properties that are characteristic of ketones but can be reduced, by weak reducing agents such as sulphurous acid, to the corresponding dihydroxybenzene. Numerous biological compounds are known in which 'quinonoid' systems occur, and some of these are known to be involved in important oxidation or reduction steps in biochemical processes. In these cases a quinonoid structural feature is found, which can be reduced to the corresponding 'benzenoid' structure by an easily reversible process:

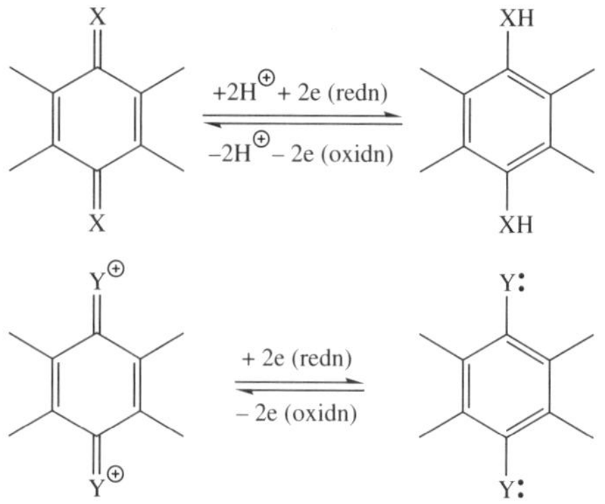

96   Carbonyl compounds: aldehydes and ketones

Complex quinones are widely distributed in nature, many as pigments:

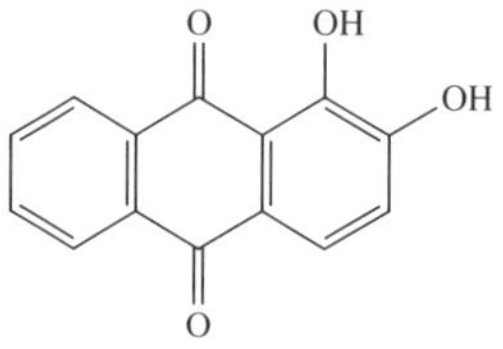

Alizarin (a plant pigment once used as a dye)

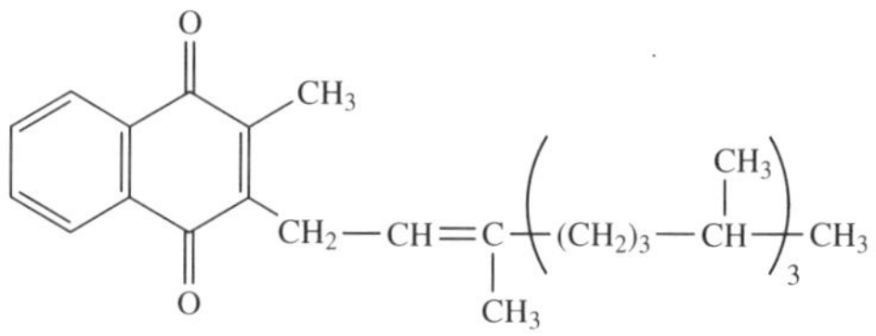

Vitamin K₁ (a blood coagulation factor)

3.3   Summary

1. The carbonyl carbon is sp² hybridised and forms a strong σ bond (sp² orbital on C overlaps with p orbital on O) and a weak π bond (p orbital on C overlaps sideways with p orbital on O). The bonds are polarised owing to the greater electronegativity of oxygen so that the carbon has a partial positive charge (δ+) and the oxygen a partial negative charge (δ–). Nucleophiles (Nu⁻) thus react at carbon, and electrophiles (E⁺) react at oxygen.

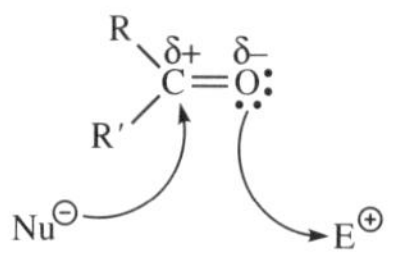

2. Aldehydes and ketones can be prepared by the oxidation of primary (1°) and secondary (2°) alcohols, respectively. The oxidation of a primary alcohol requires a mild reducing agent since aldehydes themselves can be oxidised to carboxylic acids:

$$R-CH_2OH \longrightarrow [R-CHO] \longrightarrow R-CO_2H$$

$$R-CH(OH)-R' \longrightarrow R-CO-R'$$

Another general method for the preparation of aldehydes and ketones is the oxidative cleavage of alkenes (using ozone then zinc and ethanoic acid or potassium permanganate then periodic acid):

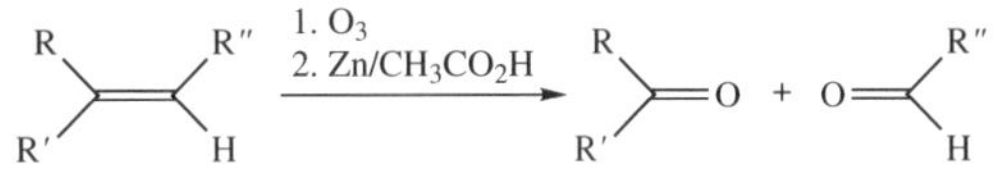

$$\underset{R'}{\overset{R}{>}}=\underset{H}{\overset{R''}{<}} \quad \xrightarrow[\text{2. Zn/CH}_3\text{CO}_2\text{H}]{\text{1. O}_3} \quad \underset{R'}{\overset{R}{>}}=O \;+\; O=\underset{H}{\overset{R''}{<}}$$

Aromatic aldehydes can also be prepared via the hydrolysis of the readily available *gem*-dihaloalkanes, whilst aromatic ketones can be prepared by Friedel–Crafts acylation of an aromatic ring (acid chloride in the presence of a Lewis acid):

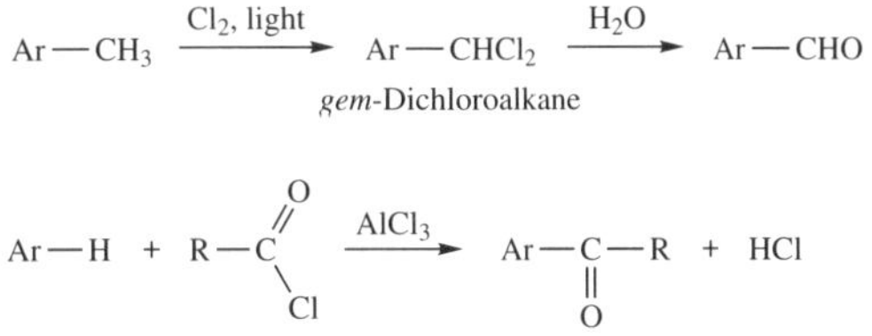

$$\text{Ar}-\text{CH}_3 \xrightarrow{\text{Cl}_2,\text{ light}} \underset{\textit{gem}\text{-Dichloroalkane}}{\text{Ar}-\text{CHCl}_2} \xrightarrow{\text{H}_2\text{O}} \text{Ar}-\text{CHO}$$

$$\text{Ar}-\text{H} \;+\; \text{R}-\overset{\overset{O}{\|}}{\underset{Cl}{C}} \xrightarrow{\text{AlCl}_3} \text{Ar}-\underset{\underset{O}{\|}}{C}-\text{R} \;+\; \text{HCl}$$

3. Aldehydes and ketones readily undergo nucleophilic addition reactions, with a variety of nucleophiles. The product of the initial addition depends upon the nucleophile, and can be a simple adduct (e.g. reaction with H^-, HCN, H_2O, ROH and $HSO_3^-Na^+$) or the product of a condensation reaction (e.g. reaction with NH_3 and its derivatives).

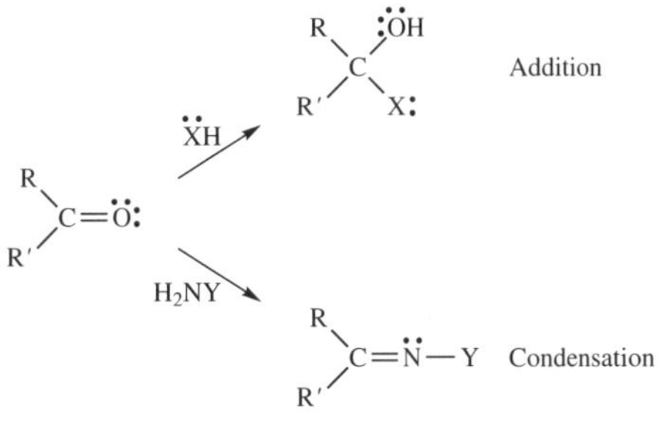

Additions can be either acid- or base-catalysed (when the nucleophile is weak), or require no catalysis (strong nucleophile). Acid catalysis involves protonation of the carbonyl group to form a more electrophilic species, which is more susceptible to nucleophilic attack.

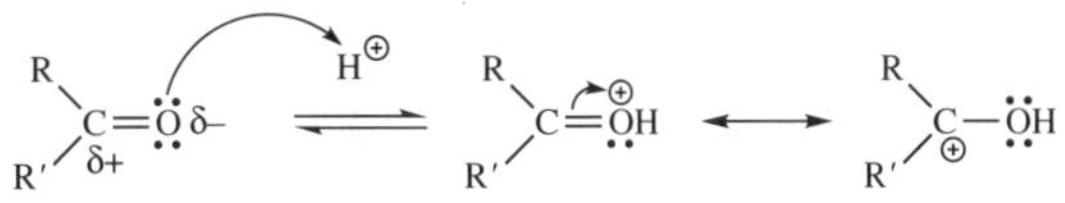

4. Aldehydes or ketones with α-hydrogens are in equilibrium with their enol tautomers. The interconversion of keto and enol tautomers can be either acid- or base-catalysed.

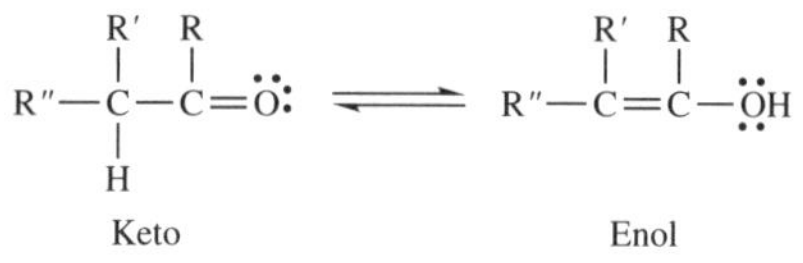

$$R''-\underset{\underset{H}{|}}{\overset{\overset{R'}{|}}{C}}-\overset{\overset{R}{|}}{C}=\ddot{O}: \rightleftharpoons R''-\overset{\overset{R'}{|}}{C}=\overset{\overset{R}{|}}{C}-\ddot{\ddot{O}}H$$

         Keto   Enol

Under acid- or base-catalysed conditions such aldehydes and ketones can dimerise via the aldol addition:

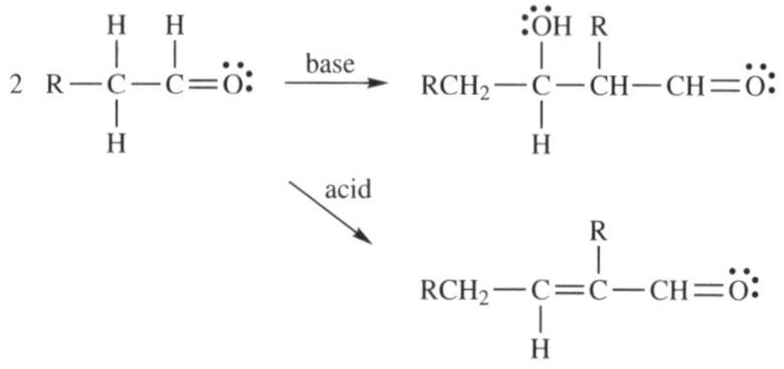

$$2\ R-\underset{\underset{H}{|}}{\overset{\overset{H}{|}}{C}}-\overset{\overset{H}{|}}{C}=\ddot{O}: \xrightarrow{\text{base}} RCH_2-\underset{\underset{H}{|}}{\overset{\overset{\ddot{O}H}{|}}{C}}-\overset{\overset{R}{|}}{CH}-CH=\ddot{O}:$$

$$\xrightarrow{\text{acid}} RCH_2-\underset{\underset{H}{|}}{C}=\overset{\overset{R}{|}}{C}-CH=\ddot{O}:$$

5. Another example of a carbon–carbon bond-forming reaction is the Grignard synthesis of alcohols, in which a carbon nucleophile, RMgX (the Grignard reagent), reacts with an aldehyde or ketone to give an alcohol:

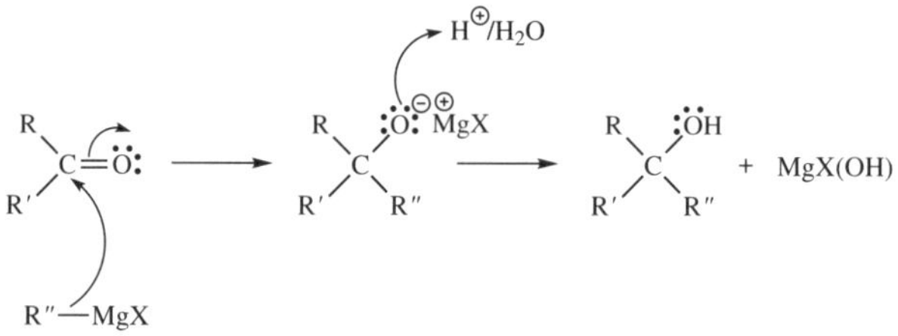

$$\underset{R'}{\overset{R}{\diagdown}}C=\ddot{O}: \longrightarrow \underset{R'\ \ R''}{\overset{R\ \ :\ddot{O}:^{\ominus}\ MgX^{\oplus}}{\diagup\diagdown}}C \xrightarrow{H^{\oplus}/H_2O} \underset{R'\ \ R''}{\overset{R\ \ :\ddot{O}H}{\diagup\diagdown}}C\ +\ MgX(OH)$$

$$R''-MgX$$

Problems

3.1 For the reaction scheme below, provide the reagents, **a**–**e**, and the unknown compounds (**X**)–(**Z**).

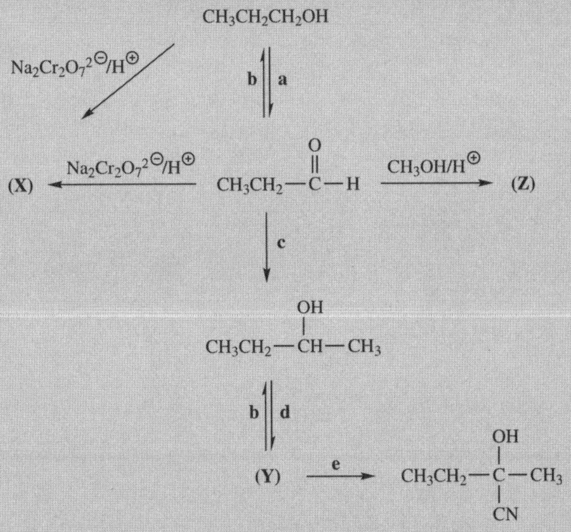

$$CH_3CH_2CH_2OH$$

$$Na_2Cr_2O_7{}^{2\ominus}/H^{\oplus} \text{ (b↑↓a)}$$

$$(X) \xleftarrow{Na_2Cr_2O_7{}^{2\ominus}/H^{\oplus}} CH_3CH_2-\underset{\underset{\|}{O}}{C}-H \xrightarrow{CH_3OH/H^{\oplus}} (Z)$$

↓ c

$$CH_3CH_2-\underset{\underset{|}{OH}}{CH}-CH_3$$

b ↑↓ d

$$(Y) \xrightarrow{e} CH_3CH_2-\underset{\underset{|}{CN}}{\overset{\overset{|}{OH}}{C}}-CH_3$$

3.2 A solution of acetone (propanone) in $H_2{}^{18}O$ slowly forms $(CH_3)_2C^{18}O$. What is the mechanism of this reaction? Do you expect the reaction to be accelerated by acid or base?

3.3 *Nonylphenol* is widely used as a plasticiser. One of the isomers present in the nonylphenol mixture is shown below. How could you prepare this nonylphenol isomer starting from phenol and using reactions you have met in this chapter?

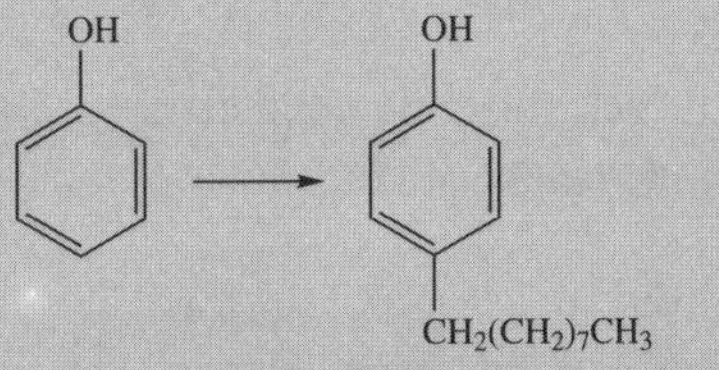

Phenol → Nonylphenol (one isomer) [para-substituted with $CH_2(CH_2)_7CH_3$]

3.4 Acetone (propanone) reacts with phenol in the presence of an acid catalyst to give *bisphenol A*, a plasticiser (which has recently been shown to exhibit oestrogen-like properties). Suggest a mechanism for this reaction.

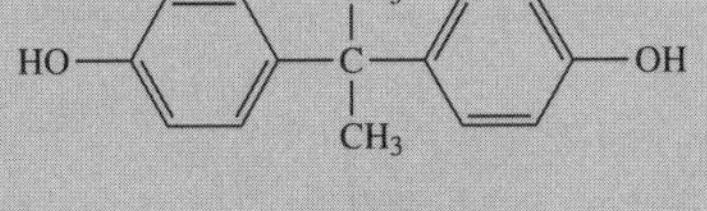

HO—C$_6$H$_4$—C(CH$_3$)$_2$—C$_6$H$_4$—OH   Bisphenol A

3.5 By what mechanism are *glyceraldehyde* and *dihydroxypropanone* interconverted? Would the reaction be catalysed by acid or base?

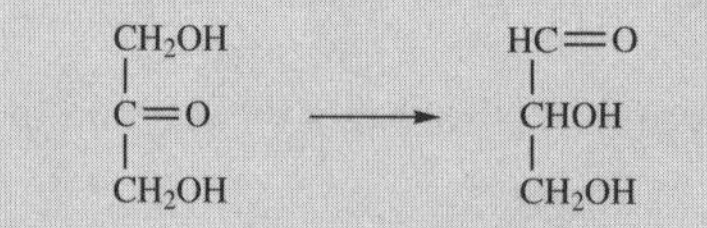

100   Carbonyl compounds: aldehydes and ketones

$$\underset{\text{Dihydroxypropanone}}{\begin{array}{c} CH_2OH \\ | \\ C=O \\ | \\ CH_2OH \end{array}} \longrightarrow \underset{\text{Glyceraldehyde}}{\begin{array}{c} HC=O \\ | \\ CHOH \\ | \\ CH_2OH \end{array}}$$

3.6 Give reagents and conditions for the following conversions:

(i) 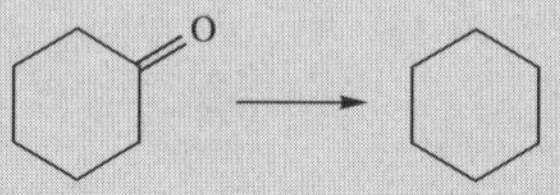 cyclohexanone → cyclohexane

(ii) 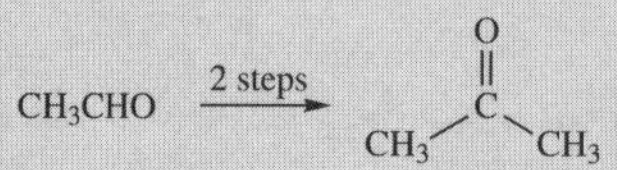 $CH_3CHO \xrightarrow{\text{2 steps}} CH_3-CO-CH_3$

(iii)  $CH_3-CO-CH_3 \longrightarrow$ 2,2-dimethyl-1,3-dithiolane (CH₃)₂C(S–CH₂–CH₂–S)

(iv) 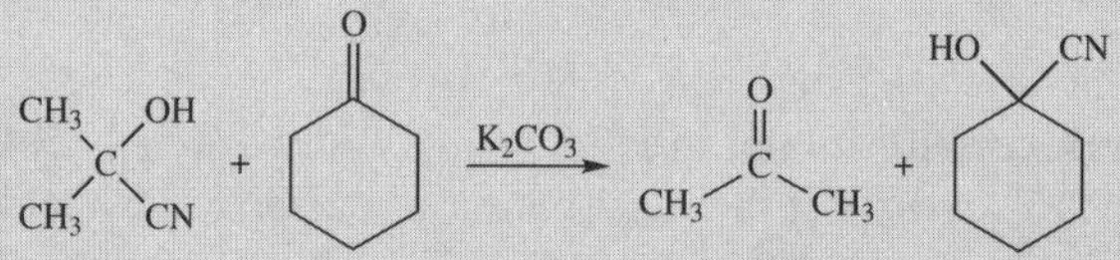 $(CH_3)_2C(OH)(CN) + \text{cyclohexanone} \xrightarrow{K_2CO_3} CH_3-CO-CH_3 + \text{1-hydroxycyclohexanecarbonitrile}$

3.7 Describe additions to carbonyl groups in aldehydes and ketones, including examples of formation of bonds from H, O, N and C atoms.

3.8 Copy the following mechanistic scheme and complete it by adding all the missing charges and curly arrows (to represent the movement of electrons).

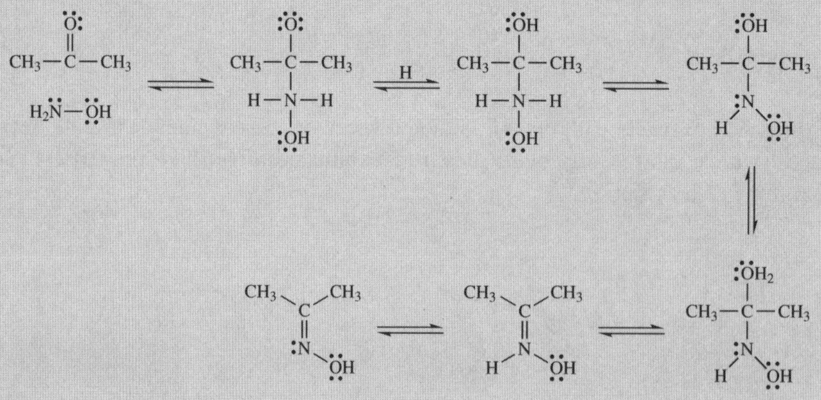

[Mechanism: acetone + H₂N–OH → tetrahedral intermediate → proton transfers → oxime formation via dehydration]

3.9 Draw all the possible enol tautomers of the following species:

(i) $CH_3CH_2-\overset{O}{\underset{\|}{C}}-CH_3$

(ii) $(CH_3)_3C-\overset{O}{\underset{\|}{C}}-C(CH_3)_3$

(iii) 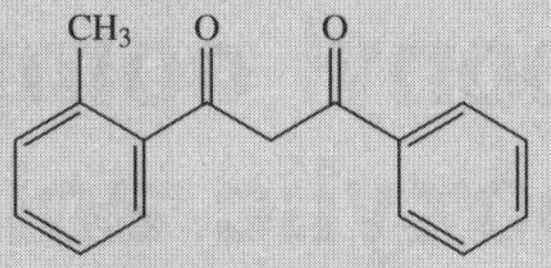 2-methylphenyl-CO-CH$_2$-CO-phenyl

3.10 The classic test for methyl ketones is the **iodoform reaction**:

$$CH_3-\overset{O}{\underset{\|}{C}}-R \xrightarrow{OH^{\ominus}, I_2} \left[CI_3-\overset{O}{\underset{\|}{C}}-R\right] \longrightarrow RCO_2^{\ominus} + HCI_3$$
$$\text{(A)} \qquad\qquad\qquad \text{Iodoform}$$

Suggest a mechanism for the formation of the intermediate (A).
[Hint: the methyl group, in this case, has three α-hydrogens.]

4 Carbonyl compounds: carboxylic acids and their derivatives

Topics

4.1 Carboxylic acids
4.2 Carboxylic acid derivatives: esters
4.3 Carboxylic acid derivatives: acyl halides
4.4 Carboxylic acid derivatives: acid anhydrides
4.5 Carboxylic acid derivatives: thioesters
4.6 Carboxylic acid derivatives: amides
4.7 Summary

Carboxylic acids contain the **carboxyl** functional group

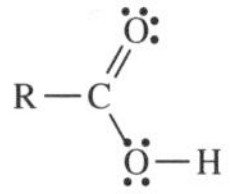

$$R-C(=\ddot{O})-\ddot{O}-H$$

The carboxylic acids, as the name implies, exhibit marked acidity owing to the relative ease with which they lose a proton (see p. 105). The properties of the carboxyl group in aliphatic and aromatic carboxylic acids are essentially the same but the nature of the R group does influence some of the properties, e.g. acid strength.

4.1 Carboxylic acids

4.1.1 Nomenclature

The IUPAC system of nomenclature indicates the presence of a carboxyl group by the suffix '**-oic acid**' attached to the name of the parent alkane. The longest chain carrying the —CO$_2$H is said to be the parent structure and is named by replacing the **-e** of the corresponding alkane by **-oic acid**. Substituents are given the lowest possible number but the carboxyl carbon is always C-1.

CH$_3$(CH$_2$)$_8$CO$_2$H CH$_3$(CH$_2$)$_3$—CH(C$_6$H$_5$)—CH(OH)—CO$_2$H

Decanoic acid   2-Hydroxy-3-phenylheptanoic acid

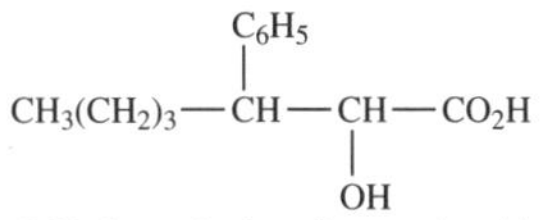

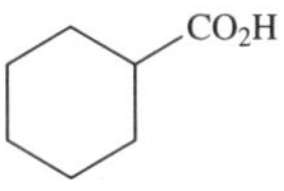

Cyclohexanecarboxylic acid

The alternative system of nomenclature regards the carboxyl group as a substituent and adds the suffix '**carboxylic acid**' to the name of the corresponding alkane.

As with many other classes of organic compounds, long-established, trivial names persist in use for the first few members of the series (Table 4.1) and,

Table 4.1

	Trivial name	Systematic name
HCO$_2$H	Formic acid (occurs in sting of an ant bite)	Methanoic acid
CH$_3$CO$_2$H	Acetic acid (produced by bacterial oxidation of ethanol; occurs naturally, e.g. in vinegar)	Ethanoic acid
CH$_3$CH$_2$CO$_2$H	Propionic acid	Propanoic acid
CH$_3$CH$_2$CH$_2$CO$_2$H	Butyric acid (gives rancid butter its smell)	Butanoic acid
CH$_3$CH$_2$CH$_2$CH$_2$CO$_2$H	Valeric acid	Pentanoic acid

once again, the designation of the carbon atoms of the chain by Greek letters is still widely used:

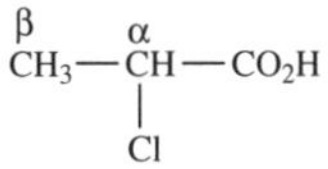

$\overset{\beta}{\text{CH}_3}$—$\overset{\alpha}{\text{CH}}$(Cl)—CO$_2$H   2-Chloropropanoic acid
(α-chloropropionic acid)

The anions derived from carboxylic acids are described by the suffixes '**-oate**' or '**-carboxylate**'.

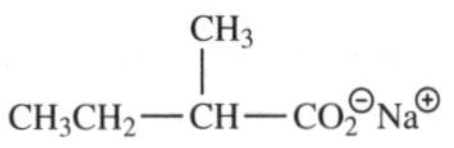

CH$_3$CH$_2$—CH(CH$_3$)—CO$_2^{\ominus}$Na$^{\oplus}$ Potassium cyclobutane-CO$_2^{\ominus}$K$^{\oplus}$

Sodium 2-methylbutanoate 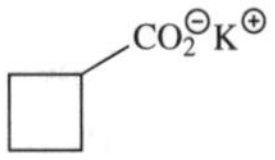   Potassium cyclobutane-carboxylate

4.1.2 Preparation

Oxidation of primary alcohols and aldehydes

Carboxylic acids may be prepared by the strong oxidation of a primary alcohol (see p. 65) or by the oxidation of an aldehyde (see p. 84):

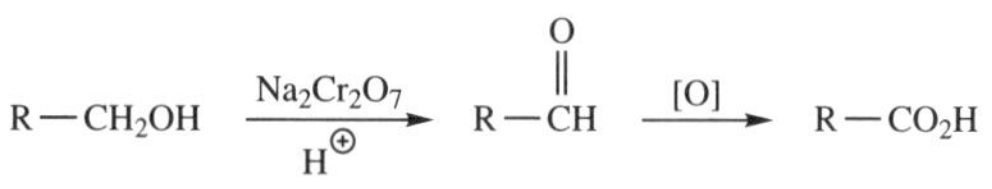

$$R-CH_2OH \xrightarrow[H^{\oplus}]{Na_2Cr_2O_7} R-\overset{O}{\overset{\|}{C}}H \xrightarrow{[O]} R-CO_2H$$

Hydrolysis of carboxylic acid derivatives

The hydrolysis of esters (p. 113), anhydrides, acyl chlorides, amides (p. 118) or nitriles leads to the corresponding carboxylic acids:

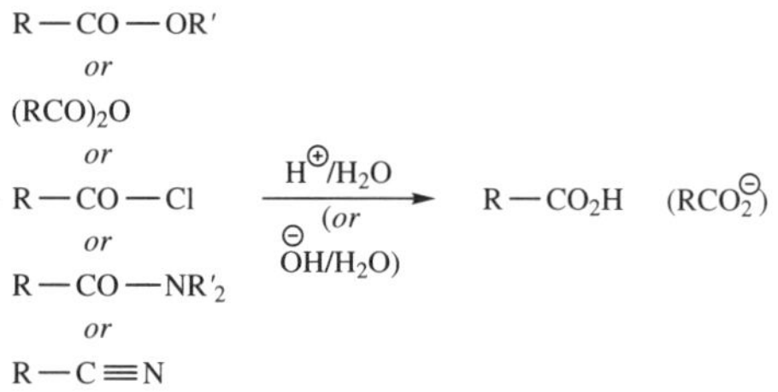

$$\begin{matrix} R-CO-OR' \\ \text{or} \\ (RCO)_2O \\ \text{or} \\ R-CO-Cl \\ \text{or} \\ R-CO-NR'_2 \\ \text{or} \\ R-C\equiv N \end{matrix} \xrightarrow[\overset{\ominus}{OH}/H_2O)]{H^{\oplus}/H_2O} R-CO_2H \quad (RCO_2^{\ominus})$$

Oxidation of alkylbenzenes

Aromatic carboxylic acids can be prepared by the strong oxidation of alkylbenzenes:

$$R-C_6H_4-CH_3 \xrightarrow[\underset{\overset{\ominus}{OH}}{KMnO_4}]{\overset{Na_2Cr_2O_7}{H^{\oplus}} \text{ or}} R-C_6H_4-CO_2H$$

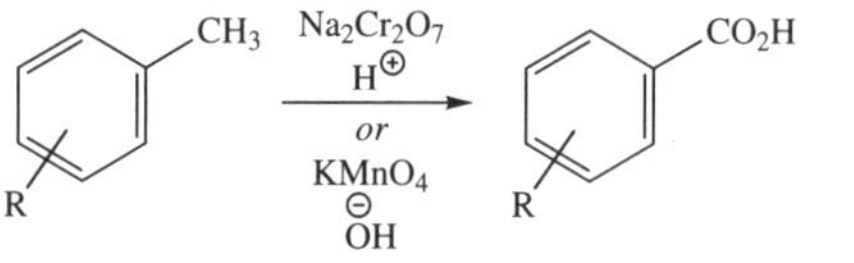

Carbonation of Grignard reagents

As mentioned previously, Grignard reagents are carbon nucleophiles and so react with electrophilic centres. If carbon dioxide is reacted with a Grignard reagent, a carboxylate anion is formed, which can subsequently be protonated to give the carboxylic acid:

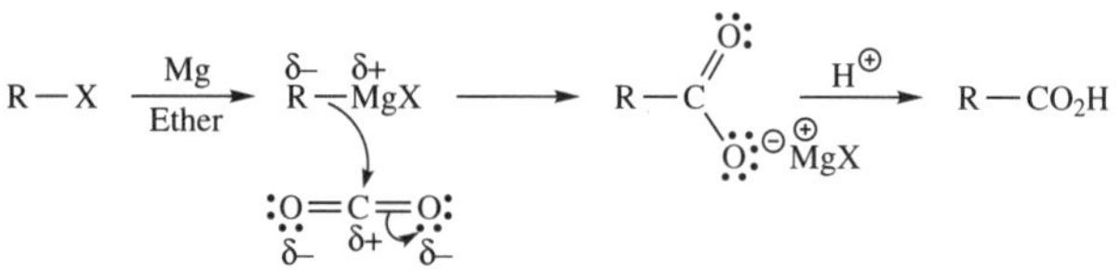

$$R-X \xrightarrow[\text{Ether}]{Mg} \overset{\delta-}{R}-\overset{\delta+}{MgX} \longrightarrow R-C\overset{\overset{\ddot{O}:}{\|}}{\underset{\ddot{O}:\overset{\ominus}{\ }\overset{\oplus}{MgX}}{\ }} \xrightarrow{H^{\oplus}} R-CO_2H$$

$$\underset{\delta-\ \ \delta+\ \ \delta-}{:\ddot{O}=C=\ddot{O}:}$$

4.1.3 Properties

The lower aliphatic carboxylic acids are pungent liquids of higher boiling points than the corresponding alcohols and are miscible with, or very soluble in, water. These properties can be attributed to the polarity of the carboxyl group and the ability of carboxylic acid molecules to form hydrogen bonds with each other, and with other molecules. Thus, the solubility in water is due to the formation of hydrogen bonds with water molecules while the greater boiling points (than those of the alcohols) are due to dimerisation, with the dimer held together by two hydrogen bonds:

$$R-C\begin{matrix}\ddot{O}:\cdots\cdots H-\ddot{O}\\ \\ \ddot{O}-H\cdots\cdots:\ddot{O}\end{matrix}C-R$$

The solubility of the simple carboxylic acids decreases progressively as the non-polar hydrocarbon chain increases in size, with the long-chain fatty acids (e.g. stearic acid, $CH_3(CH_2)_{16}CO_2H$) being waxy solids, insoluble in water. The aromatic carboxylic acids contain too many carbon atoms to be appreciably soluble in water.

4.1.4 Acidity and acid strength (pH and p*K*)

As mentioned previously, the carboxylic acids are acidic. They are much weaker acids than the mineral acids but much stronger than the alcohols and water.

Acid–base systems are of the greatest importance in biological reactions, and it is necessary to be able to describe both the acidity (or alkalinity) of solutions and the relative strengths of acids and bases by a convenient quantitative scale. pH and p*K* are the functions used for this purpose.

pH

Pure water dissociates to a minute extent, $H_2O \rightleftarrows HO^- + H^+$, and the equilibrium constant for this dissociation is given by the expression:

$$K_e(H_2O) = \frac{[H^+][OH^-]}{[H_2O]}$$

where $[H^+]$, etc., are the concentrations in $mol\,dm^{-3}$. Since in pure water and dilute aqueous solutions $[H_2O]$ is very much larger than $[H^+]$ or $[OH^-]$, and virtually constant, the expression can be rewritten as:

$$K_e(H_2O)[H_2O] = [H^+][OH^-] = \text{ionic product}$$

and for pure water and dilute aqueous solutions at 25 °C, $[H^+][OH^-] = 10^{-14}\,(mol\,dm^{-3})^2$.

Thus, for pure water $[H^+] = [OH^-] = 10^{-7}\,mol\,dm^{-3}$, and in dilute solutions of acid or alkali the ratio of $[H^+]$ and $[OH^-]$ is fixed by the ionic product.

In 1 M strong monobasic acid (i.e. fully dissociated) $[H^+] = 1$ and $[OH^-] = 10^{-14}\,mol\,dm^{-3}$, while in 1 M strong monoacid base $[OH^-] = 1$ and therefore $[H^+] = 10^{-14}\,mol\,dm^{-3}$. Many reactions occur between these limits, and the description of acidity must cover values of $[H^+]$ stretching over 14 powers of 10. In order to contract this enormous scale into workable figures the concentration of hydrogen ion is expressed as a logarithmic term, pH, which is defined by:

$$pH = -\log_{10} [H^+]$$

As $[H^+]$ changes from 1 to 10^{-14} mol dm^{-3}, the pH changes from 0 to 14. Thus, for a 0.1 M solution of a strong monobasic acid $[H^+] = 0.1$ mol dm^{-3} and so, $pH = -\log_{10}(0.1) = 1$. Similarly, 0.001 M strong monobasic acid has a pH = 3, while for a 0.01 M solution of a strong monoacid base, $[OH^-] = 0.01$, thus, $[H^+] = 10^{-12}$ mol dm^{-3} and pH = 12. The pH of pure water is 7, this being synonymous with neutrality.

(Below 10^{-6} M solutions of strong monobasic acids the concentration of ions produced by the dissociation of water becomes important. For solutions of strong monobasic acids greater than 10^{-6} M this is negligible.)

pK

The equilibrium constant for the dissociation of an acid

$$HA + H_2O \rightleftarrows H^+ + A^-$$

is given by

$$K_a = \frac{[H^+][A^-]}{[HA][H_2O]}$$

but, once again, $[H_2O]$ is virtually constant, so that the expression can be rewritten as

$$K_a = \frac{[H^+][A^-]}{[HA]}$$

and the stronger the acid, the greater the degree of dissociation, and the larger the value of K_a. The relative strengths of acids can be compared by use of K_a values, but it is frequently more convenient to use the term pK_a, which is defined by:

$$pK_a = -\log_{10} K_a = -\log_{10}\left\{\frac{[H^+][A^-]}{[HA]}\right\}$$

Polybasic acids, with several dissociation steps, have pK_1, pK_2, etc., corresponding to their first, second, etc., ionisations.

The relationships between pH and pK can be seen by considering the situation when the acid HA is exactly half dissociated. In these circumstances $[HA] = [A^-]$ so that $K_{HA} = [H^+]$ and

$$-\log_{10} K_{HA} = -\log_{10} [H^+]$$

i.e. pK_a = pH.

Thus the physical significance of the pK value for an acid is that it is the pH at which the acid is half ionised. Since pK, like pH, is a logarithmic term, a difference of unity means a tenfold difference in K_{HA}, e.g. an acid of pK = 2.5 is 10 times stronger than one of pK = 3.5 and 100 times stronger than one of pK = 4.5. The strong mineral acids, which are completely ionised even in fairly concentrated solutions, have infinitely large K_{HA} values, but the weak acids commonly encountered in organic compounds have pK values between 0 and 14.

The pK values for a series of acids, phenols, alcohols and thiols are listed in Table 4.2.

Carboxylic acids 107

Table 4.2
pK_a values of some acids

	Acid	Conjugate base*	pK_a
1	HCO_2H	HCO_2^-	3.75
2	CH_3CO_2H	$CH_3CO_2^-$	4.75
3	$CH_3CH_2CO_2H$	$CH_3CH_2CO_2^-$	4.87
4	$CH_3(CH_2)_2CO_2H$	$CH_3(CH_2)_2CO_2^-$	4.81
5	$(CH_3)_2CHCO_2H$	$(CH_3)_2CHCO_2^-$	4.84
6	$ClCH_2CO_2H$	$ClCH_2CO_2^-$	2.85
7	Cl_2CHCO_2H	$Cl_2CHCO_2^-$	1.48
8	Cl_3CCO_2H	$Cl_3CCO_2^-$	0.70
9	$CH_3CHClCO_2H$	$CH_3CHClCO_2^-$	2.83
10	$CH_2ClCH_2CO_2H$	$CH_2ClCH_2CO_2^-$	3.98
11	$CH_3CH_2CHClCO_2H$	$CH_3CH_2CHClCO_2^-$	2.86
12	$CH_3CHClCH_2CO_2H$	$CH_3CHClCH_2CO_2^-$	4.05
13	$CH_2Cl(CH_2)_2CO_2H$	$CH_2Cl(CH_2)_2CO_2^-$	4.52
14	$HOCH_2CO_2H$	$HOCH_2CO_2^-$	3.12
15	$CH_3CH(OH)CO_2H$	$CH_3CH(OH)CO_2^-$	3.83
16	$C_6H_5CO_2H$	$C_6H_5CO_2^-$	4.19
17	$2\text{-}NO_2C_6H_4CO_2H$	$2\text{-}NO_2C_6H_4CO_2^-$	2.16
18	$3\text{-}NO_2C_6H_4CO_2H$	$3\text{-}NO_2C_6H_4CO_2^-$	3.47
19	$4\text{-}NO_2C_6H_4CO_2H$	$4\text{-}NO_2C_6H_4CO_2^-$	3.41
20	$2\text{-}ClC_6H_4CO_2H$	$2\text{-}ClC_6H_4CO_2^-$	2.92
21	$3\text{-}ClC_6H_4CO_2H$	$3\text{-}ClC_6H_4CO_2^-$	3.82
22	$4\text{-}ClC_6H_4CO_2H$	$4\text{-}ClC_6H_4CO_2^-$	3.98
23	CH_3OH	CH_3O^-	ca. 18
24	C_2H_5OH	$C_2H_5O^-$	ca. 18
25	Glycerol	Monoanion	14.15
26	C_2H_5SH	$C_2H_5S^-$	10.64
27	$CH_3(CH_2)_2SH$	$CH_3(CH_2)_2S^-$	10.83
28	C_6H_5OH	$C_6H_5O^-$	9.89
29	C_6H_5SH	$C_6H_5S^-$	7.47
30	NH_4^+	NH_3	9.25
31	$CH_3\overset{+}{N}H_3$	CH_3NH_2	10.64
32	$(CH_3)_2\overset{+}{N}H_2$	$(CH_3)_2NH$	10.72
33	$(CH_3)_3\overset{+}{N}H$	$(CH_3)_3N$	9.74
34	$C_2H_5\overset{+}{N}H_3$	$C_2H_5NH_2$	10.75
35	$(C_2H_5)_2\overset{+}{N}H_2$	$(C_2H_5)_2NH$	10.98
36	$(C_2H_5)_3\overset{+}{N}H$	$(C_2H_5)_3N$	10.76
37	$C_6H_5CH_2\overset{+}{N}H_3$	$C_6H_5CH_2NH_2$	9.37
38	$C_6H_5\overset{+}{N}H_3$	$C_6H_5NH_2$	4.58
39	$(C_6H_5)_2\overset{+}{N}H_2$	$(C_6H_5)_2NH$	0.88
40		$(C_6H_5)_3N$	Non-basic
41	$C_6H_5\overset{+}{N}H_2CH_3$	$C_6H_5NHCH_3$	4.70
42	$C_6H_5\overset{+}{N}H(CH_3)_2$	$C_6H_5N(CH_3)_2$	5.16
43	$2\text{-}NO_2C_6H_4NH_3^+$	$2\text{-}NO_2C_6H_4NH_2$	−0.26
44	$3\text{-}NO_2C_6H_4NH_3^+$	$3\text{-}NO_2C_6H_4NH_2$	2.5
45	$4\text{-}NO_2C_6H_4NH_3^+$	$4\text{-}NO_2C_6H_4NH_2$	1.0
46	H_2O	HO^-	15.75
47	C_2H_2	C_2H^-	ca. 22
48	NH_3	NH_2^-	ca. 35
49	$C_6H_5NH_2$	$C_6H_5\bar{N}H$	ca. 27
50	CH_3COCH_3	$CH_3CO\bar{C}H_2$	ca. 20
51	C_2H_4	$C_2H_3^-$	ca. 40
52	C_2H_6	$C_2H_5^-$	>40

*The anion of an acid is termed the 'conjugate base' of the acid, and the protonated cation of a base is known as the 'conjugate acid' of the base.

Note the following trends:

1. The large increase in pK_a on going from methanoic acid to ethanoic acid, and the relatively small change thereafter on extension of the side chain (1–5).
2. The large decrease in pK_a with progressive substitution of the methyl group of ethanoic acid by chlorine (2, 6, 7, 8).
3. The rapidly diminishing effect of a chlorine substituent in the side chain of an aliphatic acid as the position of substitution moves away from the carboxylic acid group (9–13).
4. The (inductive) effect of electronegative substituents on the pK_a of benzoic acid, and the variation of this effect with the position of substitution (16–22).
5. The comparative acidity of thiol and hydroxyl groups in monofunctional compounds (23–29).
6. The comparative basicities of alkylamines, arylamines and ammonia (the more strongly basic the amine, the less acidic the protonated cation) (30–42).

pK of bases

There are two common ways of describing and comparing the strengths of bases on a logarithmic scale, differing in the criterion used to measure basicity. The strength of a base is inversely related to the degree of dissociation of the protonated cation (the 'conjugate acid') derived from the base. If a base B gives rise to a cation (BH)$^+$, then this cation can be regarded as a weak acid:

$$(BH)^+ \rightleftarrows B + H^+$$

for which a dissociation constant can be written

$$K_a = \frac{[B][H^+]}{[(BH)^+]}$$

thus

$$pK_a = -\log_{10} K_a = -\log_{10}\left\{\frac{[B][H^+]}{[(BH)^+]}\right\}$$

Note the similarity of this expression and that of the pK of an acid. The stronger a base, the smaller the K_a and so the larger the pK_a.

Alternatively, the strength of the base can be related to the concentration of hydroxide anions generated in a dilute aqueous solution of the base.

$$B + H_2O \rightleftarrows (BH)^+ + OH^-$$

for which

$$K_b = \frac{[(BH)^+][OH^-]}{[B]}$$

if [H₂O] is regarded as a constant.
We therefore define another logarithmic scale:

$$pK_b = -\log_{10} K_b = -\log_{10}\left\{\frac{[(BH)^+][OH^-]}{[B]}\right\}$$

and, since $[H^+][OH^-] = 10^{-14}$, $pK_b = 14 - pK_a$.

Of the two scales the former is the more convenient, as the pK_a for a base, which is numerically equal to the pH at which the base is half in the protonated form (i.e. the cation is half dissociated), is directly comparable with the pK scale for acids, whereas pK_b is not.

The pK_a values of some common bases are listed in Table 4.2. Note the general increase in basic strength (increase in pK_a) with progressive substitution of the hydrogen atoms of ammonia by alkyl groups, and the very marked decrease in basicity when the substituent is an aromatic group.

4.1.5 Reactions

Salt formation

As we have already said, the most characteristic reaction of carboxylic acids is their ionisation:

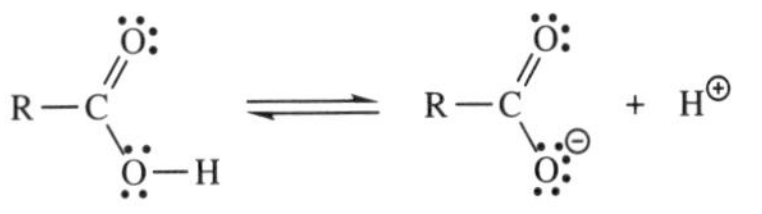

$$R-C(=O)-O-H \rightleftharpoons R-C(=O)-O^- + H^+$$

The relative ease with which this ionisation occurs is attributable to two causes:

1. The displacement of electrons along the double bond of the carbonyl group towards the oxygen atom leaves a partial positive charge on the carbon atom, which causes an inductive displacement along the C—OH and O—H bonds away from the hydrogen atom. The hydrogen atom is thus electron-deficient and can be removed by interaction with a base (the hydrogen is acidic). Ionisation of carboxylic acids is appreciable only in the presence of suitable proton acceptors, e.g. H₂O, and is negligible in hydrocarbon solvents.

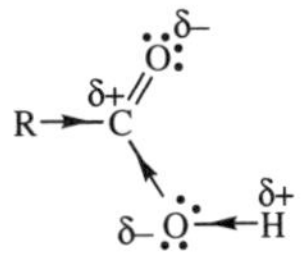

2. The anion produced by loss of a proton is a resonance hybrid of two canonical structures. The delocalisation of the charge stabilises the anion ('conjugate base'), which is therefore more easily formed, e.g. in aqueous solution

$$HA + H_2O \rightleftharpoons H_3O^\oplus + A^\ominus$$

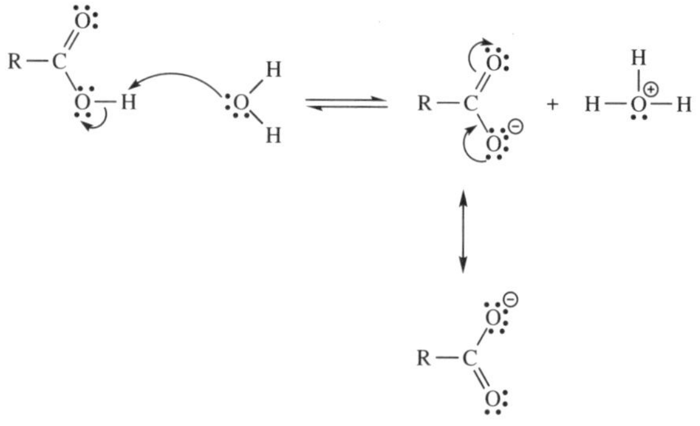

[Mechanism diagram showing carboxylic acid protonation equilibrium with water, forming protonated carboxylic acid and H_3O^+, with resonance structure of carboxylate anion shown below]

The simple aliphatic acids are weak acids giving rise to stable salts with strong bases, e.g. sodium acetate, $CH_3CO_2^-Na^+$. They are stronger acids than carbonic acid and will, therefore, liberate carbon dioxide from carbonates and bicarbonates, this being part of a simple test for the presence of a carboxyl group:

$$RCO_2H + NaHCO_3 \longrightarrow RCO_2^\ominus Na^\oplus + H_2O + CO_2\uparrow$$

Conversion into derivatives

Carboxylic esters. In the presence of mineral acids, or very slowly in their absence, carboxylic acids react with alcohols to form esters:

$$RCO_2H + HOR' \underset{}{\overset{H^\oplus}{\rightleftharpoons}} \underset{\text{Ester}}{R-\overset{\overset{O}{\|}}{C}-OR'} + H_2O$$

e.g. $\underset{\text{Ethanoic acid}}{CH_3CO_2H} + C_2H_5OH \overset{H^\oplus}{\rightleftharpoons} \underset{\substack{\text{Ethyl acetate} \\ \text{(ethyl ethanoate)}}}{CH_3-\overset{\overset{O}{\|}}{C}-OC_2H_5} + H_2O$

This reaction is reversible under the reaction conditions employed so that equilibrium mixtures are obtained. The method of choice for the preparation of esters is frequently via the acid chloride (see later). Esters can also be formed by the interaction of silver salts of carboxylic acids with alkyl halides in a typical nucleophilic displacement:

$$RCO_2^\ominus Ag^\oplus + R'I \longrightarrow RCO_2R' + AgI$$

Amides. When the ammonium salts of carboxylic acids are heated, amides are formed slowly,

$$RCO_2^\ominus \overset{\oplus}{N}H_4 \xrightarrow{\text{Heat}} R-\overset{\overset{O}{\|}}{C}-NH_2 + H_2O$$

The direct conversion of carboxylic acids into amides can be achieved by heating a mixture of the ammonium salt and carboxylic acid, or by passing a stream of ammonia gas continuously into the heated acid. Once again, however, the method of choice for the preparation of amides is often via the acid chloride.

$$R-CO_2H \longrightarrow R-\underset{\underset{\parallel}{O}}{C}-Cl \begin{array}{l} \xrightarrow{NH_3} R-\underset{\underset{\parallel}{O}}{C}-NH_2 + HCl \quad (1°) \\ \xrightarrow{R'NH_2} R-\underset{\underset{\parallel}{O}}{C}-NHR' + HCl \quad (2°) \\ \xrightarrow{R'NHR''} R-\underset{\underset{\parallel}{O}}{C}-N\underset{R''}{\overset{R'}{\diagup}} \; | \; HCl \quad (3°) \end{array}$$

Acyl halides (acid halides). Carboxylic acids react with non-metal halides to form acyl halides:

$$R-CO_2H \begin{array}{l} \xrightarrow{PCl_3, \, PCl_5 \, or \, SOCl_2} R-\underset{\underset{\parallel}{O}}{C}-Cl \quad \text{Acid chloride} \\ \xrightarrow{PBr_3} R-\underset{\underset{\parallel}{O}}{C}-Br \quad \text{Acid bromide} \end{array}$$

For the preparation of acid chlorides, thionyl chloride ($SOCl_2$) is most used since the by-products are gaseous (SO_2 and HCl), thus producing an irreversible reaction and allowing easy purification. This reaction proceeds via the formation of a chlorosulphite (a good leaving group):

$$R-\underset{\underset{\parallel}{O}}{C}-\ddot{O}-H \xrightarrow{-H^\oplus} R-\underset{\underset{\parallel}{O}}{C}-\ddot{O}-S-\ddot{Cl} \longrightarrow R-\underset{\underset{\parallel}{O}}{C}-Cl + SO_2 + HCl$$

All of the above three reactions formally involve only the hydroxyl group. Carboxylic acids do not show the addition and condensation reactions that are characteristic of aldehydes and ketones.

Reduction
The reduction of carboxylic acids is very difficult and the normal metal–acid reducing systems are ineffective. Lithium aluminium hydride reduces carboxylic acids smoothly to the corresponding alcohols, as does high-pressure hydrogenation.

$$R-CO_2H \xrightarrow[\substack{or \\ H_2 \, (100 \, atm; \, 1.013 \times 10^7 \, Pa)/ \\ CuCrO_2 \, catalyst}]{LiAlH_4} R-CH_2OH$$

112 Carbonyl compounds: carboxylic acids and their derivatives

4.2 Carboxylic acid derivatives: esters

4.2.1 Nomenclature

Esters are the *O*-alkyl or *O*-aryl derivatives of carboxylic acids, and their systematic names are derived either from a combination of those of the alkyl (aryl) group and the carboxylic acid, or by describing the ester group (acyl group) as a substituent:

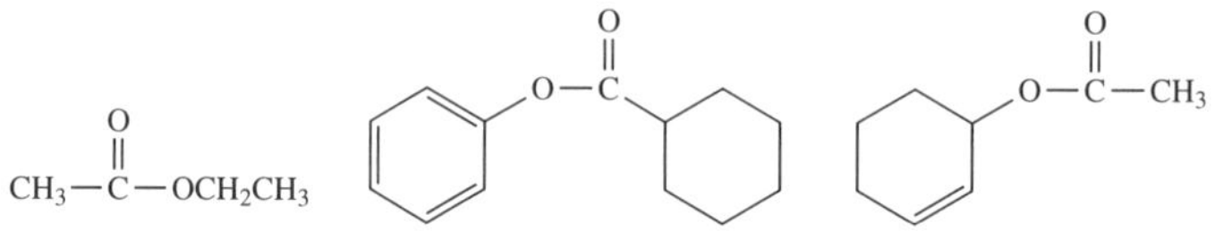

Ethyl acetate (ethyl ethanoate) — $CH_3-C(=O)-OCH_2CH_3$

Phenyl cyclohexane-carboxylate

3-Acetoxycyclohexene

4.2.2 Preparation

As mentioned previously (p. 110), esters can be prepared by the acid-catalysed reaction of carboxylic acids with alcohols:

$$R-CO_2H + R'OH \underset{}{\overset{H^\oplus}{\rightleftharpoons}} R-CO_2R' + H_2O$$

The mechanism of this reaction is:

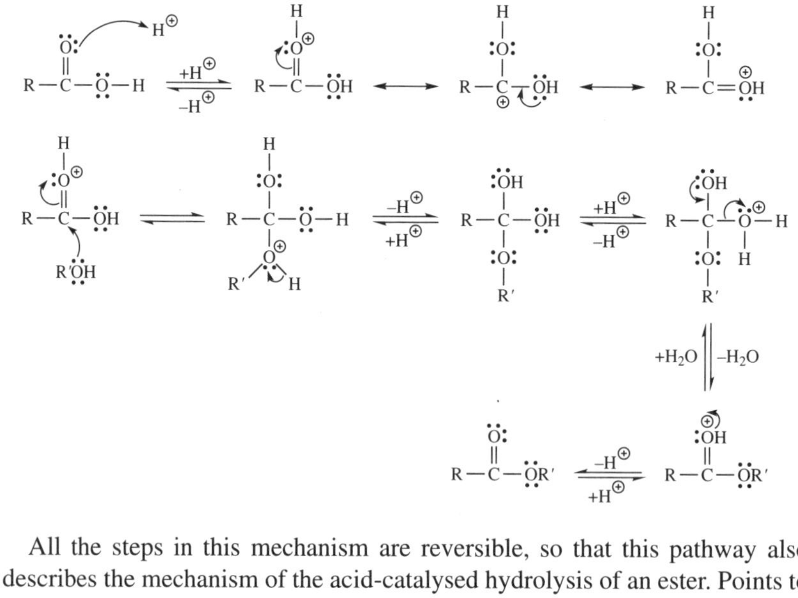

[mechanism diagram showing stepwise protonation, addition of R'OH, proton transfers, loss of water, and deprotonation to give $R-C(=O)-OR'$]

All the steps in this mechanism are reversible, so that this pathway also describes the mechanism of the acid-catalysed hydrolysis of an ester. Points to note about this mechanism are:

1. The initial protonation on the carbonyl oxygen produces a resonance-stabilised cation. As in the case of the formation of *gem*-diols, this

protonation increases the electrophilicity of the carbonyl group, making it more susceptible to attack by the poorly nucleophilic alcohol. Protonation occurs on the carbonyl oxygen as this gives rise to a more stable carbocation than that formed by protonation on the hydroxyl group.

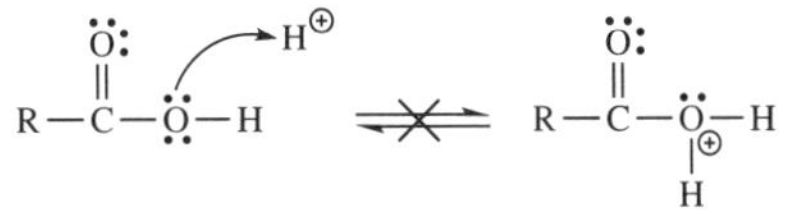

2. Protonation of one of the hydroxyl groups of the intermediate formed by attack of the alcohol converts the hydroxyl into a good leaving group (lost as water).

Esters may also be prepared by the reaction of silver salts of carboxylic acids with alkyl halides (p. 110) but the method of choice is by reaction of alcohols with acyl chlorides (p. 115) or acid anhydrides (p. 116).

$$R-CO_2H \xrightarrow[SOCl_2]{PCl_3, PCl_5 \text{ or}} R-CO-Cl \text{ (Acid chloride)} \xrightarrow{R'-OH} R-CO-OR' + HCl$$

Although this involves two steps, both are irreversible and usually high yielding. The acyl chloride is often prepared *in situ*, or used crude, and, if thionyl chloride is used to prepare the acyl chloride, the by-products are all gaseous.

4.2.3 Reactions

Hydrolysis
Esters react with water, forming the corresponding alcohol and carboxylic acid. This hydrolysis is slow under neutral conditions, but greatly accelerated by mineral acids, following the reverse of the esterification mechanism (p. 112). Ester hydrolysis is also accelerated under alkaline conditions:

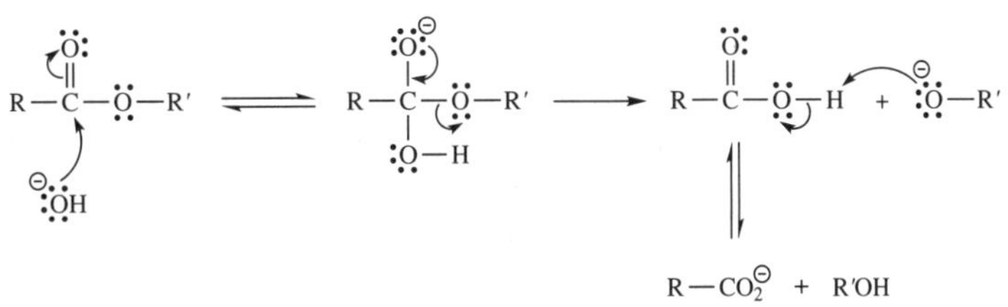

This reaction is irreversible and, in this case, the nucleophile is the strongly nucleophilic hydroxide ion, HO^-.

Transesterification
Just as an ester can undergo acidic hydrolysis, so an ester of one alcohol can undergo an acid-catalysed reaction with a second alcohol resulting in an equilibrium mixture of the two possible esters:

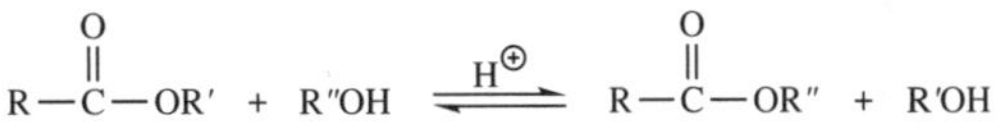

$$R-\overset{O}{\underset{\|}{C}}-OR' + R''OH \underset{}{\overset{H^\oplus}{\rightleftharpoons}} R-\overset{O}{\underset{\|}{C}}-OR'' + R'OH$$

An equilibrium mixture of esters is also produced by the reaction of an ester of one alcohol with the alkoxide anion of another:

$$R-\underset{\underset{O}{\|}}{C}-OR' + R''O^{\ominus} \rightleftharpoons R-\underset{\underset{O}{\|}}{C}-OR'' + R'O^{\ominus}$$

Reaction with ammonia

Esters react slowly with ammonia (or primary and secondary amines) to give the corresponding amides:

$$R-\underset{\underset{O}{\|}}{C}-OR' + R-NH_2 \longrightarrow R-\underset{\underset{O}{\|}}{C}-NHR + R'OH$$

A related reaction is the reaction of esters with hydroxylamine, in the presence of strong bases (e.g. alcoholic KOH), to give hydroxamic acids:

$$H_2N-OH + C_2H_5O^{\ominus} \rightleftharpoons C_2H_5OH + H\ddot{N}^{\ominus}-OH$$

$$R-\underset{\underset{O}{\|}}{C}-OR' + H\ddot{N}^{\ominus}-OH \longrightarrow R-\underset{\underset{O}{\|}}{C}-NHOH + R'O^{\ominus}$$
Hydroxamic acid

Hydroxamic acids form deep purple complexes with iron(III), and the reaction provides a colorimetric method of estimating ester groups or their equivalent (e.g. amides) in biological material such as fats.

$$R-\underset{\underset{O}{\|}}{C}-NHOH + Fe^{3+} \longrightarrow \left[R-C\underset{N-O}{\overset{\ddot{O}-Fe^{\ominus}}{\diagup}}\underset{H}{\big|}\right]^{2+} + H^{\oplus}$$

Reduction

Like carboxylic acids, esters are resistant to reduction by most of the common reducing agents, but reduction to the corresponding alcohols can be achieved by use of lithium aluminium hydride (which can be regarded as a source of hydride ions):

$$R-\underset{\underset{:H^{\ominus}}{\nearrow}}{\overset{\overset{:\ddot{O}:}{\|}}{C}}-\ddot{O}-R' \longrightarrow R-\underset{H}{\overset{:\ddot{O}:^{\ominus}}{\underset{|}{C}}}-\ddot{O}-R' \longrightarrow R-\underset{\underset{:H^{\ominus}}{\nearrow}}{\overset{\overset{:\ddot{O}:}{\|}}{C}}-H \longrightarrow RCH_2\ddot{O}:^{\ominus}$$

$$+ R\ddot{O}:^{\ominus} \qquad \Big\downarrow H^{\oplus}/H_2O$$

$$\Big\downarrow H^{\oplus}/H_2O \qquad RCH_2\ddot{O}H$$

$$R\ddot{O}H$$

The initial attack of hydride ion (and subsequent loss of the alkoxide ion) results in the formation of an aldehyde, which then reacts with a hydride ion to form a second alkoxide anion. Upon addition of acid during work-up of the reaction mixture, the alkoxides are converted into the alcohols:

$$(CH_3)_2CHCO_2CH_3 \xrightarrow[\text{2. }H_2O\ (H^\oplus)]{\text{1. LiAlH}_4/\text{ether}} (CH_3)_2CHCH_2OH + CH_3OH$$

4.3 Carboxylic acid derivatives: acyl halides

Of the acyl halides the chlorides are the most commonly encountered, and their chemistry is typical of the group. The systematic nomenclature of acyl chlorides classifies them either as derivatives of the carboxylic acid (i.e. alkanoyl chloride) or by describing the functional group as a substituent, e.g.

CH$_3$CH$_2$CH$_2$CH$_2$COCl
Pentanoyl chloride

4-Methylcyclohexanecarbonyl chloride

As mentioned previously (p. 111), acyl chlorides are prepared by the reaction of carboxylic acids with non-metal chlorides.

4.3.1 Reactions

Acyl chlorides react very readily with nucleophiles, giving products similar to those obtained from the corresponding reactions of esters. The reactivity is, however, very much greater than that of the corresponding esters, since the powerfully electronegative halogen atom polarises the C—Cl bond, leaving the carbonyl carbon atom even more electrophilic than in the ester. As a result, acyl chlorides react readily with weak nucleophiles, such as water or alcohols, without acid catalysis and are, therefore, the starting materials of choice in the preparation of carboxylic acid derivatives.

$$R-\underset{\delta+}{C}(\overset{\delta-\ddot{O}:}{\|})\rightarrow\underset{\delta-}{\ddot{C}\ddot{l}:}$$

$$R-CO-Cl + \begin{cases} H_2O \\ R'OH \\ RCO_2^\ominus \\ NH_3 \\ R'NH_2 \\ R'_2NH \\ R'SH \end{cases} \longrightarrow \begin{cases} RCO_2H + HCl \\ RCO_2R' + HCl \\ (RCO)_2O + Cl^\ominus \\ RCONH_2 + NH_4^\oplus Cl^\ominus \\ RCONHR' + R'NH_3^\oplus Cl^\ominus \\ RCONR'_2 + R'_2NH_2^\oplus Cl^\ominus \\ RCOSR' + HCl \end{cases}$$

The reactions summarised above all proceed by essentially the same mechanism, exemplified below for the formation of an ester by the reaction of an acyl chloride with an alcohol:

$$R-\overset{\overset{\displaystyle :\overset{..}{O}:}{\|}}{\underset{\overset{\displaystyle)}{R'-\overset{..}{O}-H}}{C}-\overset{..}{\underset{..}{C}l}:} \; \rightleftharpoons \; R-\overset{\overset{\displaystyle :\overset{..}{O}:^{\ominus}}{|}}{\underset{\overset{\displaystyle |}{\underset{\displaystyle R'}{\overset{\displaystyle \oplus}{O}}-H}}{C}-\overset{..}{\underset{..}{C}l}:} \; \longrightarrow \; R-\overset{\overset{\displaystyle :\overset{..}{O}:}{\|}}{\underset{\overset{\displaystyle |}{\underset{\displaystyle R'}{O}}-H}{C}} \; :\overset{..}{\underset{..}{C}l}:^{\ominus} \; \rightleftharpoons \; R-CO_2R' \; + \; HCl$$

4.4 Carboxylic acid derivatives: acid anhydrides

Acid anhydrides are most easily prepared from the reaction of an acyl chloride with the sodium salt of a carboxylic acid. This method leads to mixed anhydrides if the anion and acyl chloride are derived from different acids:

$$R'-COCl \; + \; R-CO_2^{\ominus}Na^{\oplus} \; \longrightarrow \; R'-\overset{\overset{\displaystyle O}{\|}}{C}-O-\overset{\overset{\displaystyle O}{\|}}{C}-R$$

The reactivity of acid anhydrides is intermediate between that of esters and acyl chlorides. Thus, they react slowly with weak nucleophiles such as water or alcohols, but much more rapidly in the presence of acid catalysts. The range of reactions of acid anhydrides is summarised below:

$$\begin{Bmatrix} R-\overset{\overset{\displaystyle O}{\|}}{C}\diagdown \\ \diagup O \\ R-\overset{\underset{\displaystyle O}{\|}}{C} \end{Bmatrix} + \begin{Bmatrix} H_2O \\ R'OH \\ NH_3 \\ R'NH_2 \\ R'_2NH \\ R'SH \end{Bmatrix} \longrightarrow \begin{Bmatrix} 2RCO_2H \\ RCO_2R' \; + \; RCO_2H \\ RCONH_2 \; + \; RCO_2H \\ RCONHR' \; + \; RCO_2H \\ RCONR'_2 \; + \; RCO_2H \\ RCOSR' \; + \; RCO_2H \end{Bmatrix}$$

4.5 Carboxylic acid derivatives: thioesters

Thioesters are the acyl derivatives of thiols, and contain a C—S—C linkage. They may be prepared by the acid-catalysed reaction of thiols with carboxylic acids, or by reaction of thiols with acyl chlorides or acid anhydrides:

$$CH_3CO_2H \; + \; HSC_2H_5 \; \underset{}{\overset{H^{\oplus}}{\rightleftharpoons}} \; CH_3-\overset{\overset{\displaystyle O}{\|}}{C}-SC_2H_5 \; + \; H_2O$$
Ethyl thioacetate
(ethyl thioethanoate)

The reactions of thioesters resemble those of acid anhydrides and normal esters but their reactivity is closer to the anhydrides. The importance of thioesters in biochemical reactions is partly due to their enhanced reactivity, e.g. they readily acylate amines:

$$R-CO-SR' \; + \; R''NH_2 \; \longrightarrow \; R-CO-NHR'' \; + \; HSR'$$

For example, carboxylic acids are frequently incorporated into metabolic processes in cells via the formation of their esters of coenzyme A, a complex thiol which, nevertheless, exhibits the reactions of a simple thiol:

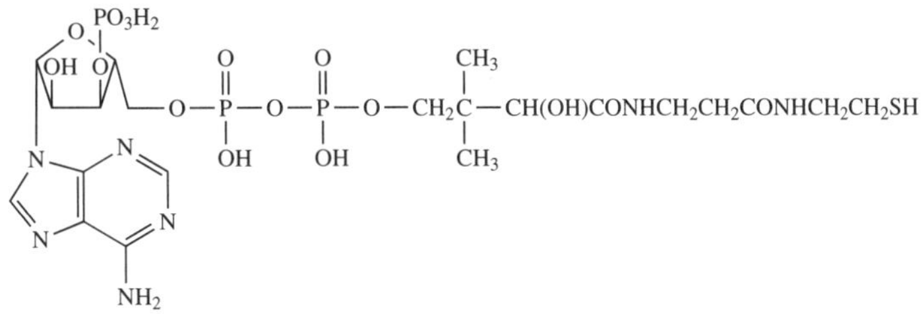

Coenzyme A (CoASH)
(Adenine-D-ribose-(phosphate)-phosphate-phosphate-pantothenic acid-mercaptoethylamine)

4.6 Carboxylic acid derivatives: amides

Amides, the acyl derivatives of amines, can be named either as derivatives of the corresponding acid or by use of 'aminocarbonyl-' to describe the functional group:

CH$_3$(CH$_2$)$_4$CONH$_2$ Cyclobutanecarboxamide
Hexanamide   (aminocarbonylcyclobutane)

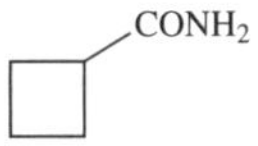

There are three types of amide, known as primary, secondary and tertiary, depending on the extent of substitution on the nitrogen atom:

RCONH$_2$ RCONHR′   RCONR′$_2$
primary (1°) secondary (2°) tertiary (3°)

Most of the preparations and reactions are common to all groups, but primary amides also exhibit a few special reactions. The method of choice for the preparation of amides is the action of ammonia or amines on a carboxylic acid derivative, preferably an acid anhydride or an acyl chloride.

4.6.1 Reactions

Salt formation
Although they contain the amino group, amides are only very feebly basic, on account of the mesomeric interaction between the carbonyl double bond and the lone-pair on the nitrogen atom:

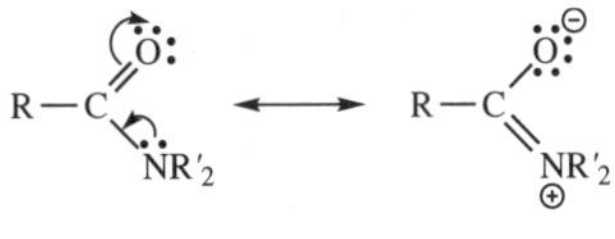

This delocalisation of the nitrogen lone-pair means that it is less available for donation (e.g. to a proton) and the basicity of amides is thus very low. The protonation of amides is thus significant only under very strongly acidic conditions and occurs on the oxygen atom, as the charge on the resultant cation

118 Carbonyl compounds: carboxylic acids and their derivatives

is delocalised, unlike the alternative nitrogen-protonated cation, in which delocalisation is impossible.

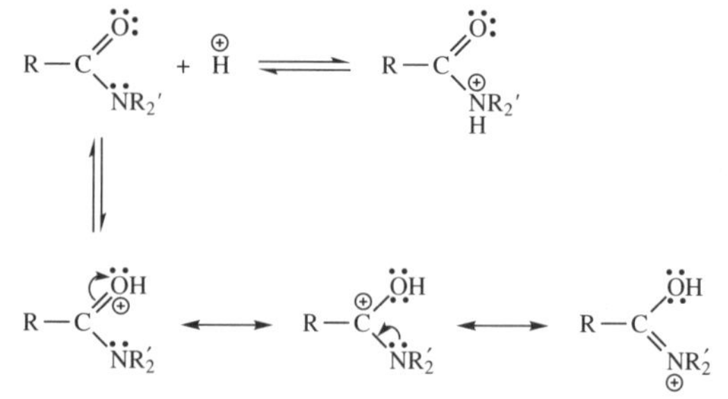

Primary and secondary amides can also act as very weak acids, losing a proton under strongly basic conditions:

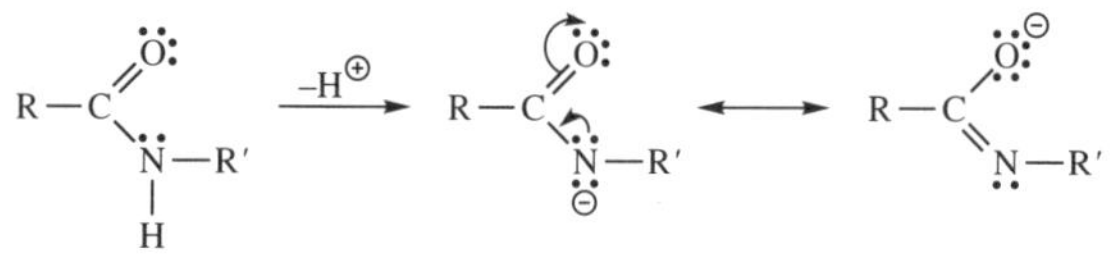

Once again, the charge on the anion is delocalised, as in the case of enolates and carboxylates.

Hydrolysis

Amides can be hydrolysed under acidic or alkaline conditions, acidic hydrolysis being the more rapid:

$$R-CO-NR'_2 + H_2O \xrightarrow{H^{\oplus}} R-CO_2H + H_2\overset{\oplus}{N}R'_2$$

$$R-CO-NR'_2 + H_2O \xrightarrow{\overset{\ominus}{O}H} R-CO_2^{\ominus} + HNR'_2$$

Dehydration

Primary amides can be dehydrated by heating with phosphorus pentoxide, forming nitriles:

$$R-CONH_2 \xrightarrow[(-H_2O)]{P_2O_5} R-C\equiv N$$

4.7 Summary

1. Carboxylic acids are usually prepared by the strong oxidation of a primary alcohol (or by the oxidation of an aldehyde):

$$R-CH_2OH \xrightarrow[H^{\oplus}]{Na_2Cr_2O_7} \left[R-\overset{O}{\underset{\|}{C}}H \right] \longrightarrow R-CO_2H$$

Another useful method of preparing carboxylic acids is by the carbonation of Grignard reagents:

$$R-X \xrightarrow{Mg} RMgX \xrightarrow[2.\ H^{\oplus}]{1.\ CO_2} R-CO_2H$$

2. Carboxylic acids are acidic, being stronger acids than water and alcohols but weaker than the mineral acids. Their acidity is due to electron displacement along the C—OH and O—H bonds making the hydrogen acidic, and the fact that the anion produced is resonance-stabilised.

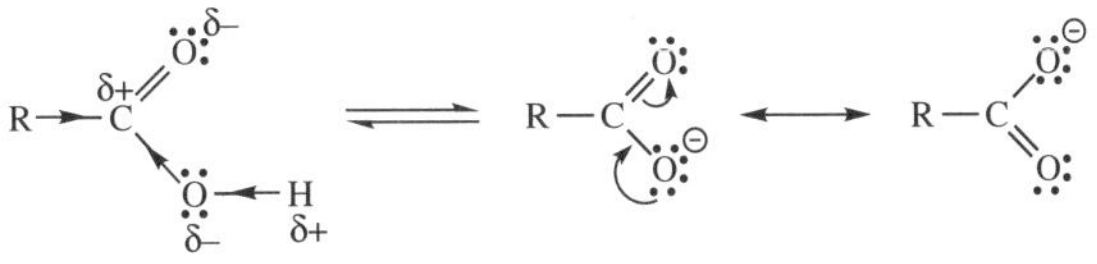

3. Carboxylic acids can be converted into a number of derivatives but this is usually easier to achieve from the acyl chlorides, which are readily prepared from the acids by reaction with an inorganic (non-metal) chloride:

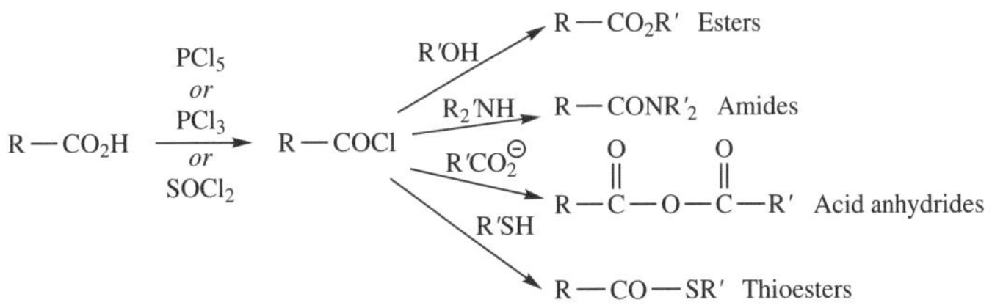

$$R-CO_2H \xrightarrow[\text{or } SOCl_2]{\text{PCl}_5 \text{ or PCl}_3} R-COCl \begin{array}{l} \xrightarrow{R'OH} R-CO_2R' \quad \text{Esters} \\ \xrightarrow{R_2'NH} R-CONR_2' \quad \text{Amides} \\ \xrightarrow{R'CO_2^{\ominus}} R-C(=O)-O-C(=O)-R' \quad \text{Acid anhydrides} \\ \xrightarrow{R'SH} R-CO-SR' \quad \text{Thioesters} \end{array}$$

4. Among the most important of the carboxylic acid derivatives are the esters (fats are naturally occurring esters of glycerol, see p. 65) and the amides (the amide group is the important link in the building blocks of life; peptides and proteins, see p. 159).

5. The interconversion of carboxylic acid derivatives is easily achieved and the order of reactivity is:

acyl chlorides > acid anhydrides > esters > carboxylic acids
most reactive   *least reactive*

Carboxylic acids, which have a poor leaving group (OH), require acid catalysis in order to undergo nucleophilic substitution, whereas acyl chlorides, which have a good leaving group (Cl), require no catalysis. The reactions of acid anhydrides and esters are generally slow unless catalysis (acid or base) is employed.

120 Carbonyl compounds: carboxylic acids and their derivatives

Problems

4.1 When chloroethanoic acid ($pK_a = 2.85$) is added to a buffer solution of pH 3.00, what proportion of the chloroethanoic acid remains undissociated?

4.2 If hydrolysis of the ester $CH_3CO_2C_2H_5$ with potassium hydroxide and $H_2^{18}O$ is interrupted before the reaction has gone to completion, the recovered ester is found to have incorporated some of the ^{18}O isotope. By considering the mechanism for this reaction, explain this observation.

4.3 By what mechanisms do the following reactions occur?

(a) The acid-catalysed reaction

$$CH_3CH_2CO_2C_6H_5 + CH_3OH \longrightarrow CH_3CH_2CO_2CH_3 + C_6H_5OH$$

(b) The base-catalysed reaction

$$(CH_3)_2NH + CH_3CO_2CH_2CH_3 \longrightarrow CH_3CON(CH_3)_2 + C_2H_5OH$$

4.4 Which of the following statements is correct?

(a) The smaller the value of pK_a, the weaker the acid.

(b) Trichloroethanoic acid ($Cl_3C.CO_2H$) is a stronger acid than ethanoic acid.

(c) Carboxylic acids are less acidic than alcohols.

(d) Electron-donating groups stabilise the conjugate bases (A^-) of acids (HA).

(e) Carboxylic acids are stronger acids than the mineral acids.

4.5 For the scheme below give the missing reagents/conditions **a–d** and the missing product **(X)–(Z)**.

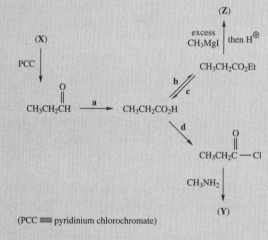

(PCC ≡ pyridinium chlorochromate)

4.6 Give reagents/conditions for the following conversions:

(a) $CH_3CHO \longrightarrow CH_3CO_2H$

(b) $CH_3CO_2H \longrightarrow CH_3CO_2CH_2C_6H_5$

(i) In *one* step
(ii) In *two* steps

(c) $CH_3CO_2H \longrightarrow CH_3CONH_2$

(i) In *one* step
(ii) In *two* steps

(d)

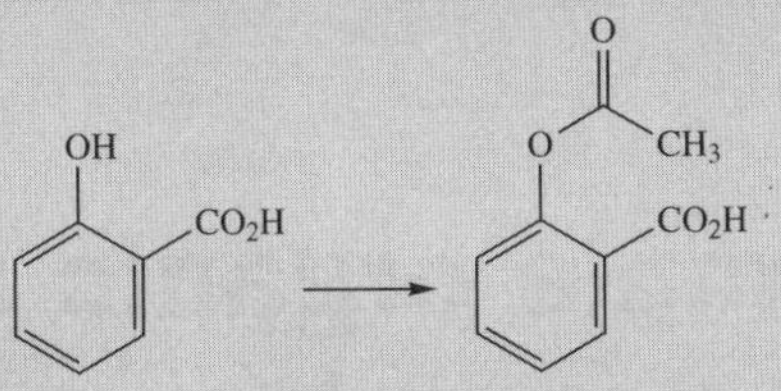

Acetylsalicylic acid
(aspirin)

4.7 By considering the stability of the resonance forms of the intermediates produced by protonation of the following species, indicate which position is most likely to be protonated:

(a) 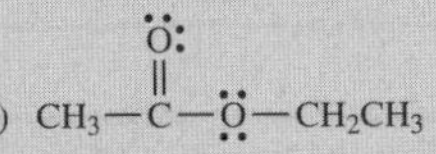

(b) 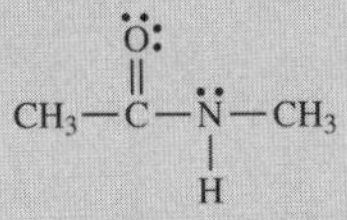

4.8 Give the complete mechanisms for the following processes:

(a) $CH_3\text{—}COCl + \overset{\ominus}{O}H \longrightarrow CH_3CO_2^{\ominus} + Cl^{\ominus}$

(b) $CH_3\text{—}CONH_2 + \overset{\oplus}{H} \longrightarrow CH_3CO_2H + \overset{\oplus}{N}H_4$

(Hint: draw all lone-pairs. Use curly arrows to represent the movement of electrons.)

5 Simple organic nitrogen compounds: amines

Topics

5.1 Amines
5.2 Aromatic diazonium salts
5.3 Summary

5.1 Amines

Amines are derived from ammonia by replacement of the hydrogen atoms of NH_3 by organic groups, and three types of amine are possible, known as primary, secondary and tertiary amines:

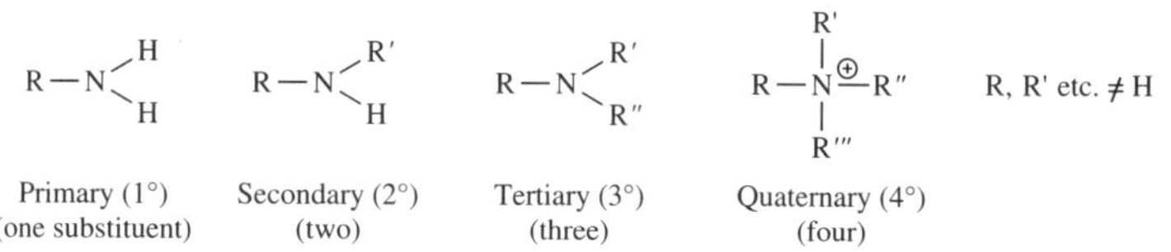

$$R-NH_2 \quad R-NHR' \quad R-NR'R'' \quad R-N^{\oplus}R'R''R''' \quad R, R' \text{ etc.} \neq H$$

Primary (1°)   Secondary (2°)   Tertiary (3°)   Quaternary (4°)
(one substituent)   (two)   (three)   (four)

In addition, a fourth class of compounds, quaternary ammonium salts, is obtained by replacement of all four of the hydrogen atoms of the ammonium cation by alkyl or aryl groups. Amines are named either by indicating the alkyl groups attached to the nitrogen followed by '**-amine**' or, for more complex amines, by using the prefix '**amino**': e.g.

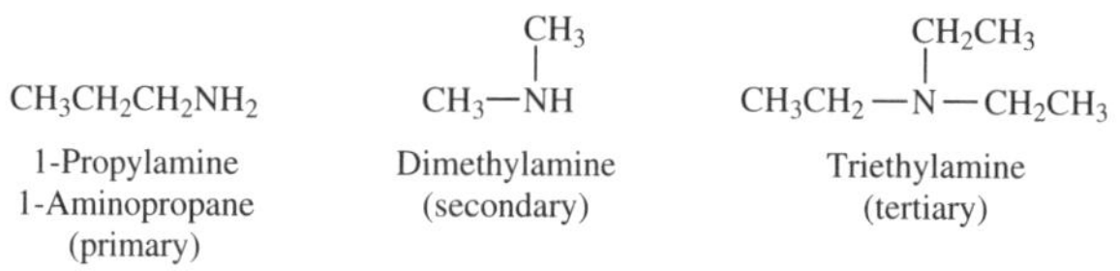

$CH_3CH_2CH_2NH_2$ $CH_3-NH-CH_3$ $CH_3CH_2-N(CH_2CH_3)-CH_2CH_3$

1-Propylamine   Dimethylamine   Triethylamine
1-Aminopropane   (secondary)   (tertiary)
(primary)

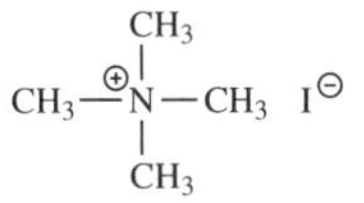

Tetramethylammonium iodide
(quaternary ammonium salt)

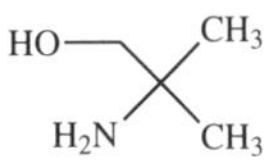

2-Amino-2-methyl-propanol
(primary)

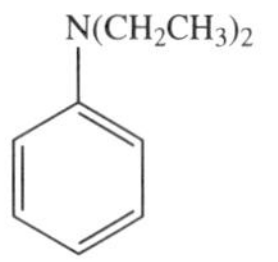

N,N-Diethylaniline
(tertiary)

5.1.1 Preparation

General

Alkylation of ammonia. The reaction of ammonia with alkyl halides gives amines and quaternary salts by a sequence of nucleophilic substitutions (p. 48). Normally a mixture of all the possible products is formed:

$$NH_3 + RBr \longrightarrow R\overset{+}{N}H_3 + Br^-$$
$$R\overset{+}{N}H_3 + NH_3 \longrightarrow RNH_2 + \overset{+}{N}H_4 \quad (1°)$$
$$R-NH_2 + RBr \longrightarrow R_2\overset{+}{N}H_2 + Br^-$$
$$R_2\overset{+}{N}H_2 + NH_3 \longrightarrow R_2NH + \overset{+}{N}H_4 \quad (2°)$$
$$R_2NH + RBr \longrightarrow R_3\overset{+}{N}H + Br^-$$
$$R_3\overset{+}{N}H + NH_3 \longrightarrow R_3N + \overset{+}{N}H_4 \quad (3°)$$
$$R_3N + RBr \longrightarrow R_4\overset{+}{N} + Br^- \quad (4° \text{ ammonium salt})$$

A large excess of ammonia favours the formation of primary and secondary amines, while an excess of alkyl halide favours the production of tertiary amines and quaternary ammonium salts. Owing to the complex mixtures obtained this method is seldom used for the preparation of amines, except aromatic amines (p. 126).

Reduction of amides. The reduction of amides by lithium aluminium hydride can give primary, secondary or tertiary amines, depending upon the amide starting material used.

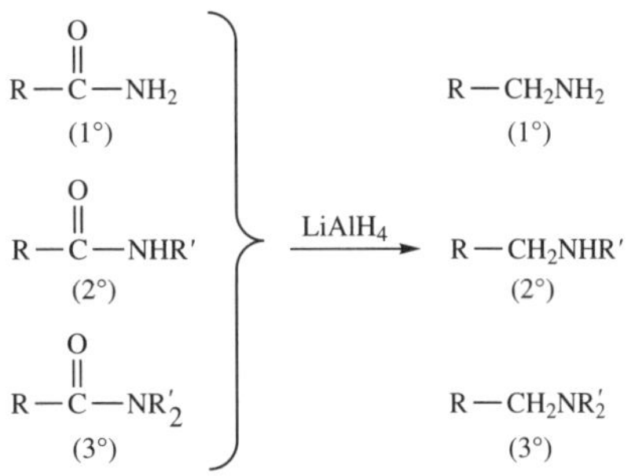

$$R-\overset{O}{\underset{\|}{C}}-NH_2 \quad (1°)$$
$$R-\overset{O}{\underset{\|}{C}}-NHR' \quad (2°) \quad \xrightarrow{LiAlH_4}$$
$$R-\overset{O}{\underset{\|}{C}}-NR'_2 \quad (3°)$$

$$R-CH_2NH_2 \quad (1°)$$
$$R-CH_2NHR' \quad (2°)$$
$$R-CH_2NR'_2 \quad (3°)$$

Decarboxylation of amino acids. This reaction is of little preparative use, but occurs widely in living systems. The enzyme-catalysed decarboxylation utilises pyridoxal phosphate as a coenzyme:

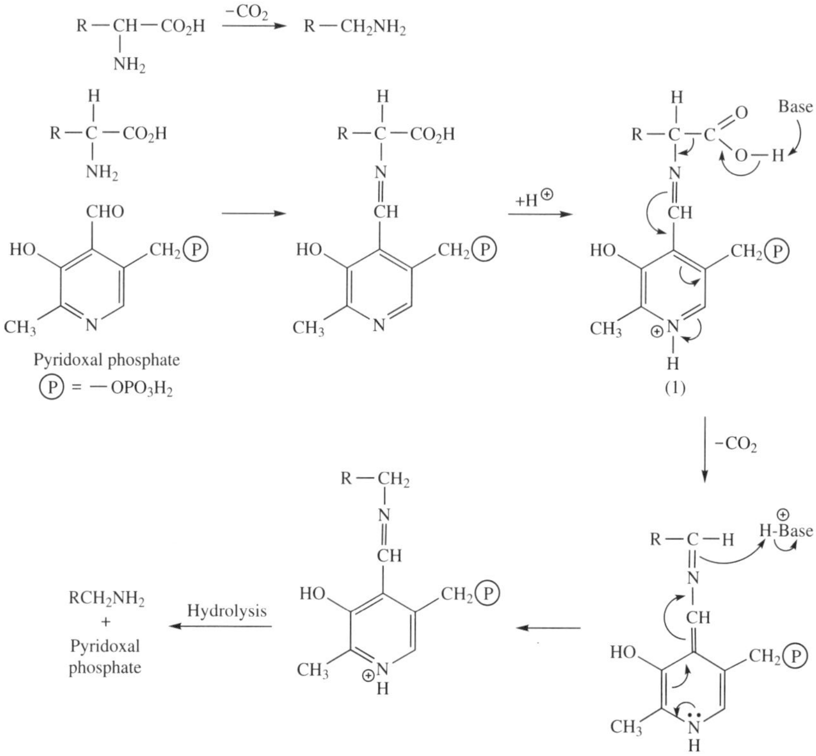

$$R-CH(NH_2)-CO_2H \xrightarrow{-CO_2} R-CH_2NH_2$$

Pyridoxal phosphate
$$\text{P} = -OPO_3H_2$$

Here the aldehyde group of pyridoxal condenses with the amino group of the amino acid to form an imine, which is then protonated. The protonated imine (**1**) has electronic features that assist decarboxylation. Finally, hydrolysis of an imine regenerates pyridoxal phosphate and gives the amine.

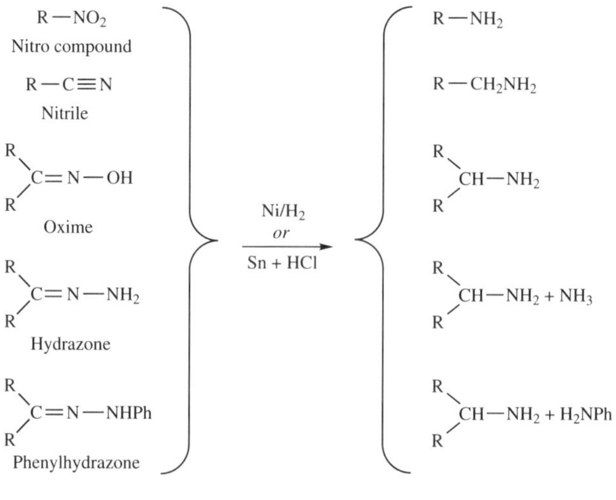

Starting material		Product
$R-NO_2$ (Nitro compound)		$R-NH_2$
$R-C\equiv N$ (Nitrile)	Ni/H$_2$ or Sn + HCl	$R-CH_2NH_2$
$R_2C=N-OH$ (Oxime)		R_2CH-NH_2
$R_2C=N-NH_2$ (Hydrazone)		$R_2CH-NH_2 + NH_3$
$R_2C=N-NHPh$ (Phenylhydrazone)		$R_2CH-NH_2 + H_2NPh$

Preparation of primary amines

Many compounds containing C—N bonds can be reduced to primary amines by catalytic methods (Ni/H$_2$) or by metal–acid reducing systems (e.g. Sn + HCl). Examples of such reductions can be seen on facing page.

The reduction of nitro compounds is particularly useful for the preparation of primary aromatic amines, since the nitro compounds can often be obtained by the direct nitration of an aromatic compound:

$$Ar-H \xrightarrow[H_2SO_4]{HNO_3} Ar-NO_2 \xrightarrow{Sn/HCl} Ar-NH_2$$

Treatment of primary amides (alkyl or aryl) (i.e. RCONH$_2$) with alkali hypobromite gives the lower primary amine (Hofmann degradation):

$$R-\overset{O}{\underset{\|}{C}}-NH_2 + KOBr \longrightarrow R-NH_2 + CO_2 + KBr$$

Preparation of secondary amines

The reduction of imines (Schiff's bases) produces secondary amines. The imines are readily obtained from aldehydes or ketones (p. 87).

$$R-CH=N-R' \xrightarrow{Pt/H_2} R-CH_2-NHR'$$

Preparation of tertiary amines

Pyrolysis or reduction of quaternary ammonium salts leads to the production of tertiary amines.

Preparation of aromatic amines

Aromatic amines can be prepared by the reaction of aryl halides (inert to nucleophiles) with ammonia at high temperatures in the presence of copper or copper salts. This is not a simple nucleophilic substitution, like the reaction of alkyl halides with ammonia, but proceeds via complex copper-containing intermediates:

$$Ar-Br + NH_3 \xrightarrow{Cu} Ar-NH_2 + NH_4Br$$

Secondary and tertiary aromatic amines can be obtained either by alkylation of primary aromatic amines by alkyl halides (nucleophilic substitution), or by reaction with aryl halides and copper at high temperatures (see above).

$$C_6H_5NH_2 \begin{cases} \xrightarrow{+CH_3I} C_6H_5NHCH_3 \xrightarrow{+CH_3I} C_6H_5N(CH_3)_2 \\ \searrow {\scriptstyle +C_6H_5I,\ Cu,\ 150°C} \\ \xrightarrow[150°C]{+C_6H_5Br,\ Cu} (C_6H_5)_2NH \xrightarrow{+CH_3I} (C_6H_5)_2NCH_3 \\ \xrightarrow[150°C]{+C_6H_5I,\ Cu} (C_6H_5)_3N \end{cases}$$

5.1.2 Properties

The lower amines are gases or liquids of characteristic ammoniacal or fishy odour, soluble in water and most organic solvents. The amines of higher relative molecular mass are progressively less water-soluble, and have repulsive odours. Aromatic amines are liquids of high boiling point, or low melting point solids, very sparingly soluble in water.

5.1.3 Reactions

Salt formation

The most characteristic property of amines is their basicity. Like ammonia, they form salts with acids, utilising the lone-pair of electrons on the nitrogen to form a bond to a hydrogen ion. Their aqueous solutions are strongly alkaline.

$$C_2H_5NH_2 + H^\oplus \rightleftharpoons C_2H_5\overset{\oplus}{N}H_3$$

$$C_2H_5NH_2 + H_2O \rightleftharpoons C_2H_5\overset{\oplus}{N}H_3 \ \overset{\ominus}{O}H$$

Aliphatic amines are stronger bases than ammonia and the increased basicity is attributable to inductive displacement along the C—N bond, making the lone-pair more available for donation. This effect also helps to stabilise the positive charge on the nitrogen atom of the cation produced, in the same way that alkyl substituents stabilise carbocations (p. 44).

$$H_3C \rightarrow \overset{..}{N}H_2 \xrightarrow{H^\oplus} \rightleftharpoons H_3C \rightarrow \overset{\oplus}{N}H_3$$

Aromatic amines are weaker bases than ammonia. Overlap of the orbital containing the lone-pair on the nitrogen atom with the aromatic π orbitals (p. 173) results in resonance between the canonical structures:

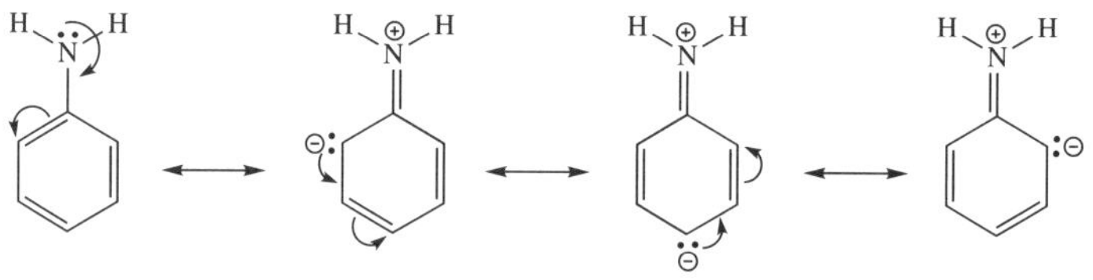

so that the lone-pair is less readily available to form a bond to a hydrogen ion than in ammonia, where no such mesomeric interaction occurs.

Amines, like ammonia, can also form complexes with metal ions by donation of the lone-pair, e.g.

$$Cu^{2+} + CH_3NH_2 \longrightarrow [Cu(NH_2CH_3)_4]^{2+} \ cf. \ [Cu(NH_3)_4]^{2+}$$

Nucleophilic attack at electrophilic centres

The lone-pair of electrons on the nitrogen atom of amines enables them to react with many compounds by nucleophilic attack at electron-deficient centres. The formation of salts (nucleophilic attack on H$^+$ of H$_3$O$^+$) and the reaction with alkyl halides (see above) are typical nucleophilic reactions. Primary and secondary amines react with derivatives of carboxylic acids (acyl chlorides, acid anhydrides, esters and thiol esters) to form amides, and these acylations are initiated by nucleophilic attack of the amine on the carbonyl group of the reagent:

$$
\begin{array}{l}
\text{R-CO-Cl (Acyl chloride)} \\
\text{R-CO-O-CO-R (Acid anhydride)} \\
\text{R-CO-OR'' (Ester)} \\
\text{R-CO-SR'' (Thioester)}
\end{array}
\Bigg\} + \text{R-NH-R'} \quad (\text{R'} = \text{alkyl, aryl, H}) \longrightarrow
\begin{array}{l}
\text{R-CO-NRR'} + \text{R'NH}_2\text{R} \overset{\oplus}{\ } \text{Cl}^{\ominus} \\
\text{R-CO-NRR'} + \text{RCO}_2\text{H} \\
\text{R-CO-NRR'} + \text{R''OH} \\
\text{R-CO-NRR'} + \text{R''SH}
\end{array}
$$

Similarly, the chlorides of sulphonic acids form sulphonamides with primary and secondary amines:

$$\text{C}_6\text{H}_5\text{SO}_2\text{Cl} + \text{R}_2\text{NH} \longrightarrow \text{C}_6\text{H}_5\text{SO}_2\text{NR}_2 + \text{R}_2\overset{\oplus}{\text{N}}\text{H}_2\,\text{Cl}^{\ominus}$$

Tertiary amines, which have no replaceable hydrogen atoms, do not form amides.

Schiff's base formation

Primary amines combine with aldehydes and ketones, eliminating water and forming imines (Schiff's bases) (p. 87):

$$\text{R-NH}_2 + \text{O=C}\begin{smallmatrix}\text{R'}\\ \text{R'}\end{smallmatrix} \longrightarrow \text{R-N=C}\begin{smallmatrix}\text{R'}\\ \text{R'}\end{smallmatrix}$$
$$\text{Imine}$$

Reaction with nitrous acid

Primary and secondary aliphatic amines react with nitrous acid, usually produced *in situ* by the addition of sodium nitrite solution to an ice-cold solution of the amine in excess dilute mineral acid. The effective reagent is the anhydride, N$_2$O$_3$, and this reacts with primary amines producing alcohols via a complex sequence of reactions:

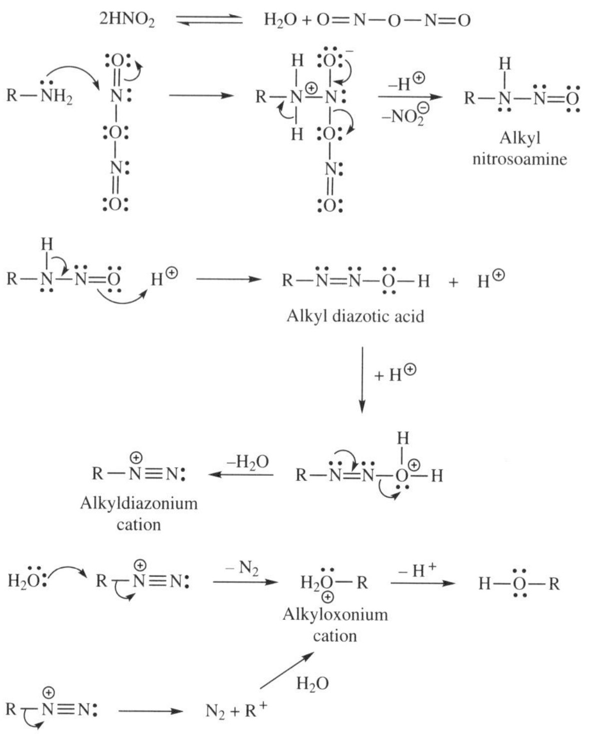

128 Simple organic nitrogen compounds: amines

$$2HNO_2 \rightleftharpoons H_2O + O=N-O-N=O$$

[Reaction scheme showing conversion of R–NH₂ through nitrosation intermediates to alkyl nitrosoamine R–N(H)–N=O]

[Scheme: alkyl nitrosoamine rearranges with H⁺ to give R–N=N–O–H (Alkyl diazotic acid) + H⁺]

[Scheme: + H⁺ gives R–N=N–O(H)–H which loses H₂O to give alkyldiazonium cation R–N≡N⁺]

[Scheme showing S_N2 attack by H₂O on R–N≡N⁺ → loss of N₂ → H₂O⁺–R (Alkyloxonium cation) → –H⁺ → H–O–R]

[Scheme showing S_N1 pathway: R–N≡N⁺ → N₂ + R⁺ → + H₂O → alkyloxonium]

Initially the primary amine is converted by the nitrous anhydride into a nitrosoamine, which, in the presence of acid, rearranges to an isomeric diazotic acid. Protonation of the oxygen atom of this intermediate by the mineral acid present, followed by elimination of water, gives an unstable alkyldiazonium ion, which rapidly decomposes to nitrogen and a carbocation. The carbocation instantly reacts with water to form an alcohol, via an alkyloxonium ion. (The conversion of a primary amine into an alcohol by nitrous acid is seldom used as a preparative method owing to the poor yield and number of by-products obtained.) This is effectively an S_N1 reaction (p. 49) of the alkyldiazonium cation. Alternatively, the diazonium ion may undergo an S_N2 reaction (p. 48) with water, the expulsion of nitrogen and formation of the alkyloxonium ion being concerted. The overall reaction is:

$$R-NH_2 + HNO_2 \longrightarrow ROH + H_2O + N_2$$

Secondary amines react with nitrous acid, but in this case the nitrosoamine produced is stable and can be isolated usually as a yellow oil or solid. Many nitrosoamines are carcinogenic:

$$R_2NH + N_2O_3 \longrightarrow R_2N-N=O + HNO_2$$
$$\text{Dialkylnitrosoamine}$$

Tertiary amines react with nitrous acid with loss of one alkyl group and formation of a nitrosoamine. The course of the reaction is thought to be:

$$R_2N-CHR'_2 \xrightarrow{N_2O_3} R_2\overset{\oplus}{N}(NO)-CHR'_2 \xrightarrow[\text{(nitroxyl)}]{-HNO} R_2\overset{\oplus}{N}=CR'_2$$

$$\downarrow H_2O$$

$$R_2N-N=O \xleftarrow{N_2O_3} R_2NH + O=CR'_2$$

Primary and second aromatic amines also react with nitrous acid; primary aromatic amines giving phenols and secondary giving nitrosoamines, but in the case of primary aromatic amines the intermediate diazonium cations are stable at 0 °C, and diazonium salts can be isolated. Tertiary aromatic amines with two alkyl groups attached to the nitrogen atom react with nitrous acid to form compounds in which the aromatic ring has been nitrosated. These nitroso-derivatives are green-coloured bases, which form orange salts.

PhN(CH$_3$)$_2$ $\xrightarrow{\text{NaNO}_2/\text{HCl}}$ [Cl$^\ominus$ H$\overset{\oplus}{N}$(CH$_3$)$_2$–C$_6$H$_4$–N=O] (orange) $\xrightarrow{\text{NaHCO}_3}$ (CH$_3$)$_2$N–C$_6$H$_4$–N=O (green)

N,*N*-Dimethylaniline → *p*-Nitroso-*N*,*N*-dimethylaniline

Oxidation

The oxidation of amines leads to a variety of products and is seldom a preparatively useful reaction. Enzymic oxidation of amines is, however, an important biological process, e.g. in the oxidative deamination of amino acids, initial dehydrogenation of the primary amino acid to an imino acid is followed by rapid hydrolysis to the corresponding ketone (an oxo-acid) and ammonia.

$$\underset{\text{Amino acid}}{R-\underset{NH_2}{\underset{|}{CH}}-CO_2H} \xrightarrow{\text{Amino acid oxidase}} R-\underset{NH}{\underset{\|}{C}}-CO_2H \xrightarrow{H_2O} \underset{\text{Oxo-acid}}{R-\underset{O}{\underset{\|}{C}}-CO_2H} + NH_3$$

In addition, the coenzyme pyridoxal phosphate (**2**) is involved in the transamination of amino acids:

$$\underset{CO_2^\ominus}{\underset{|}{R\underset{\overset{|}{\overset{\oplus}{NH_3}}}{CH}}} + R'COCO_2H \rightleftharpoons RCOCO_2H + \underset{CO_2^\ominus}{\underset{|}{R'\underset{\overset{|}{\overset{\oplus}{NH_3}}}{CH}}}$$

In the enzyme-catalysed reaction the aldehyde group of pyridoxal condenses with the amine function of the amino acid to form an imine. An acid-catalysed tautomerisation, (3) ⇌ (4), followed by hydrolysis, leads to release of the oxo-acid with formation of pyridoxamine phosphate (5).

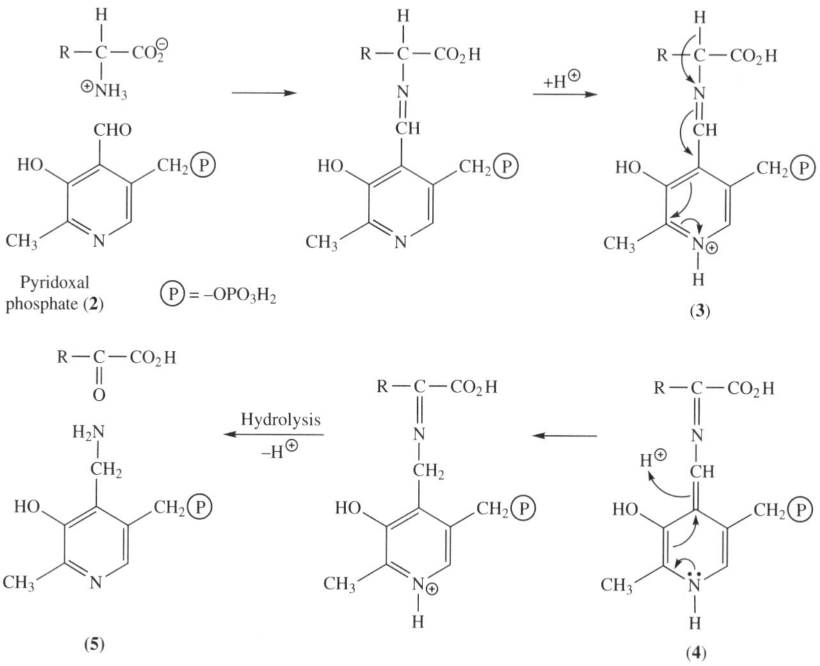

Pyridoxamine phosphate (5) can then react with a molecule of another oxo-acid by the reverse of the mechanism shown above, to regenerate pyridoxal phosphate (2) and form a new amino acid:

$$R'COCO_2H + (5) \longrightarrow (2) + R'CH(\overset{+}{N}H_3)-CO_2^-$$

Reactions of quaternary ammonium salts

Quaternary ammonium salts have few reactions. Heating the halides results in the formation of a tertiary amine and an alkyl halide, by nucleophilic attack of the halide ion on a carbon atom adjacent to the positively charged nitrogen atom – the reverse of the reaction used to prepare the salts.

$$(CH_3)_4\overset{+}{N}I^- \xrightarrow{\text{Heat}} CH_3I + (CH_3)_3N$$

Heating quaternary ammonium hydroxides (obtained by the reaction of aqueous solutions of the halides with silver oxide or by use of an ion-exchange resin) affords alkenes and tertiary amines:

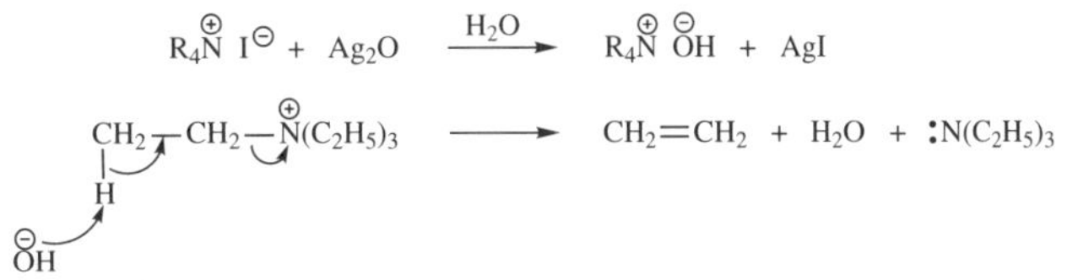

$$R_4\overset{\oplus}{N}\ I^{\ominus} + Ag_2O \xrightarrow{H_2O} R_4\overset{\oplus}{N}\ \overset{\ominus}{O}H + AgI$$

$$CH_2\text{—}CH_2\text{—}\overset{\oplus}{N}(C_2H_5)_3 \longrightarrow CH_2\text{=}CH_2 + H_2O + :N(C_2H_5)_3$$

Electrophilic substitution in arylamines

Electrophilic substitution of the aromatic ring of arylamines occurs very readily in the *ortho-* and *para-* positions. The ease of substitution is indicated by the reaction of the tertiary amines with the weakly electrophilic nitrosonium ion, NO⁺ (see above). As is the case with phenols, direct halogenation or nitration cannot be controlled and leads to polysubstituted products (phenylamine and concentrated HNO_3 react with explosive violence). However, acylation of the amino group of primary and secondary aromatic amines reduces the reactivity towards electrophiles, and mono-substituted products can be obtained by reaction of these amides with the electrophilic reagents, and subsequent hydrolysis of the amide group:

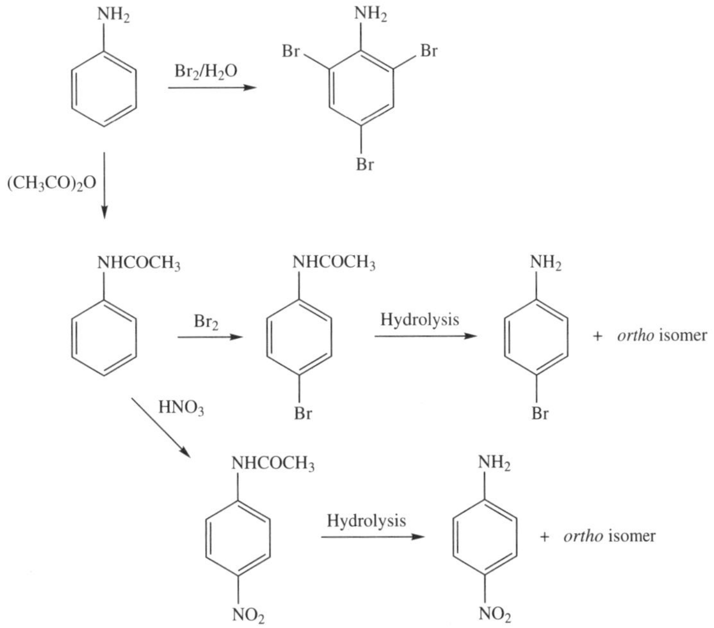

5.2 Aromatic diazonium salts

As mentioned previously, the reaction of aromatic primary amines with nitrous acid at 0 °C produces diazonium salts, which are normally stable below 5 °C. The pathway of this reaction is identical to that described for the reaction of aliphatic primary amines (p. 128):

$$C_6H_5NH_2 \xrightarrow[0°C]{NaNO_2/HCl} C_6H_5-\overset{\oplus}{N}\equiv N \;\; \overset{\ominus}{Cl}$$

These salts can be isolated, but are highly explosive crystalline solids and are usually prepared in solution and used without isolation. Although of considerable chemical significance, since the diazonium group can be readily replaced by many other functional groups, they are of little biological interest.

5.2.1 Reactions

The reactions of aromatic diazonium salts can be divided into two groups, differing in the fate of the diazonium group.

Reactions in which nitrogen is lost

The decomposition of diazonium salts in aqueous solution above 5 °C leads to phenols, with the loss of nitrogen, but in the presence of halides and cuprous salts, the diazonium group is replaced by the halogen (Sandmeyer reaction). Similarly, alkali cyanides with a cuprous cyanide catalyst will convert diazonium salts into aromatic nitriles:

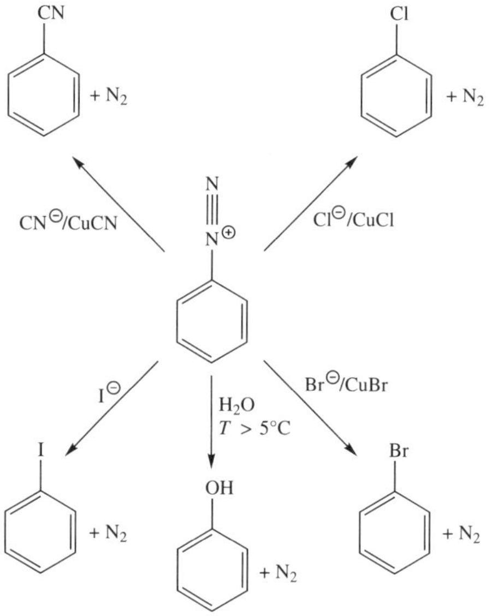

Reactions in which the diazonium group is retained

1. Reduction of diazonium salts by stannous chloride and hydrochloric acid gives arylhydrazines:

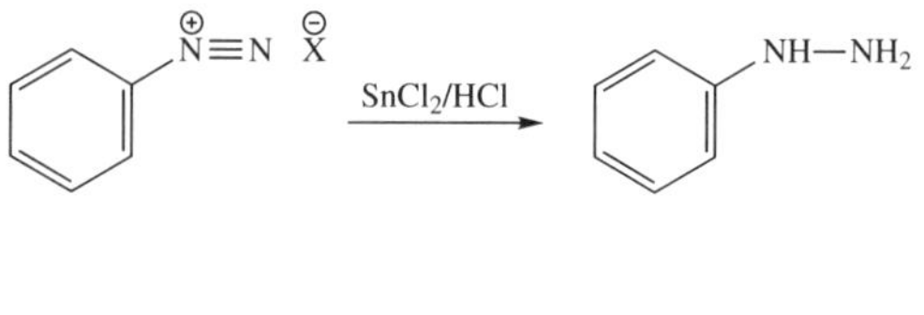

2. The diazonium cation is a weak electrophile, and will substitute reactive aromatic compounds. Phenols couple with diazonium salts in alkaline solution – the reaction is probably with the phenoxide anion – to give yellow or red 'azo-dyes' (*this being one of the tests for the presence of a primary aromatic amino functional group*). Coupling also occurs with tertiary aromatic amines, but primary and secondary aromatic amines react on the nitrogen atom, giving diaryltriazenes.

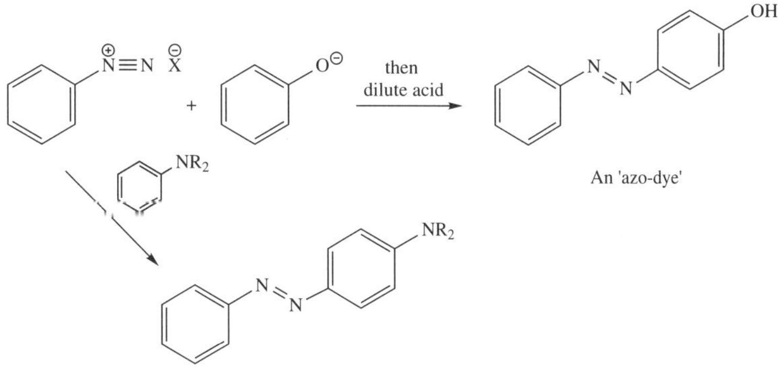

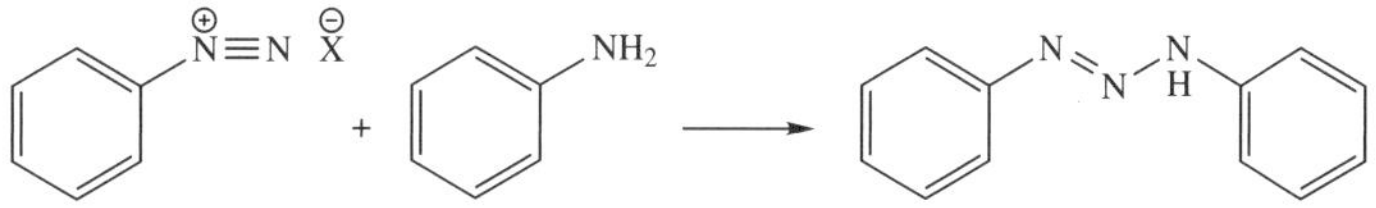

5.3 Summary

1. Amines are most readily prepared by the reduction of other functional groups. Primary, secondary and tertiary amines can be obtained by the reduction of the corresponding amides:

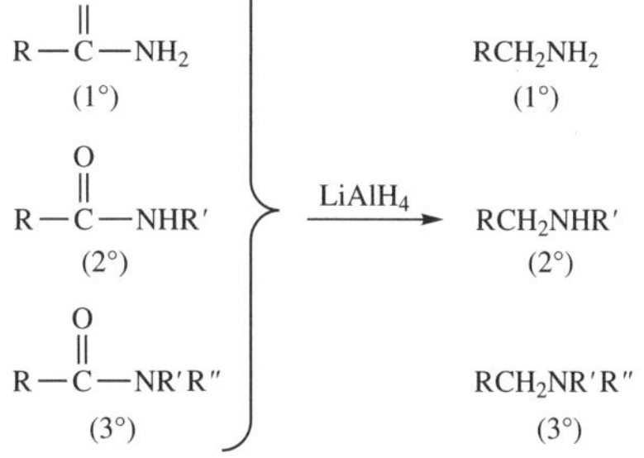

Primary amines can also be prepared by reduction of a number of compounds containing C—N bonds, with the most common being nitro compounds (particularly important for the preparation of primary aromatic amines), nitriles and oximes:

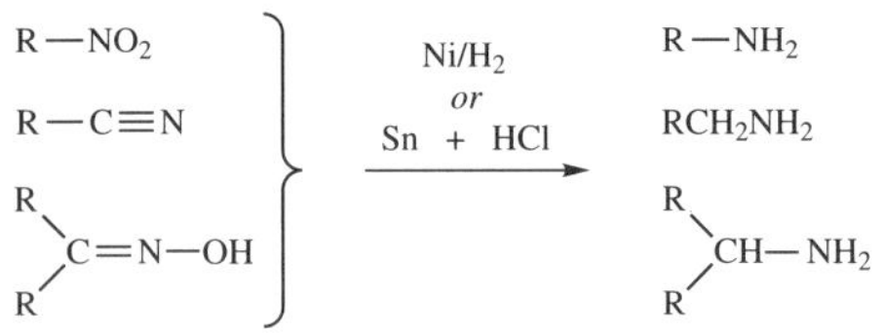

$$R-NO_2$$
$$R-C\equiv N$$
$$\begin{matrix}R\\R\end{matrix}C=N-OH$$

$\xrightarrow[\text{Sn + HCl}]{\text{Ni/H}_2 \text{ or}}$

$$R-NH_2$$
$$RCH_2NH_2$$
$$\begin{matrix}R\\R\end{matrix}CH-NH_2$$

Secondary amines can be prepared by the reduction of imines (Schiff's bases):

$$RCH=N-R' \xrightarrow{\text{Pt/H}_2} R-CH_2-NH-R'$$

2. Amines are basic and thus form salts with acids. Aliphatic amines are stronger bases than ammonia (owing to inductive displacement along the C—N bond which makes the lone-pair more available for donation and stabilises the cation formed), while aromatic amines are weaker bases than ammonia (owing to overlap of the nitrogen lone-pair with the aromatic π orbitals).

Aliphatic amines > Ammonia > Aromatic amines

($R_3N > R_2NH > RNH_2$)   ($ArNH_2 > Ar_2NH > Ar_3N$)

Decreasing basicity →

3. Amines are nucleophiles and thus react with a variety of electrophilic reagents, e.g. alkyl halides and carboxylic acid derivatives:

$$R_3N + R'Cl \longrightarrow R_3\overset{\oplus}{N}R' \; Cl^{\ominus}$$

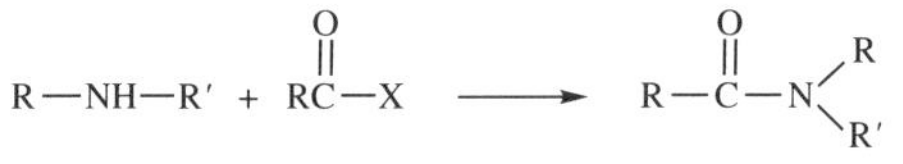

$$R-NH-R' + R\overset{\overset{O}{\|}}{C}-X \longrightarrow R-\overset{\overset{O}{\|}}{C}-N\begin{matrix}R\\R'\end{matrix}$$

(X = Cl, OCOR'', OR'', SR'')

4. Aromatic amines undergo facile electrophilic substitution in the *ortho*- and *para*- positions. Halogenation or nitration can be controlled by initial acylation to give the mono-substituted products:

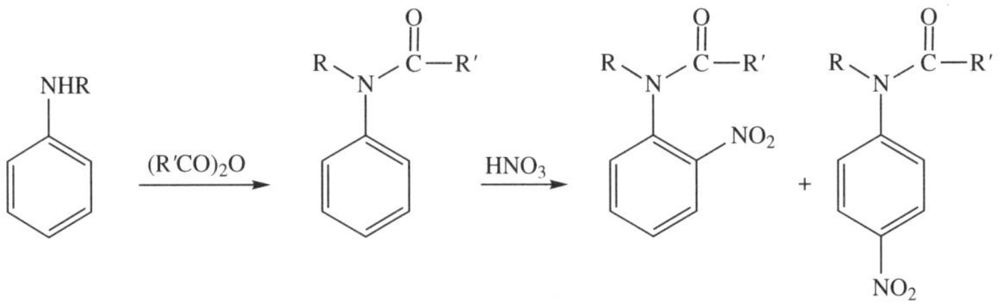

[Structures: PhNHR → (R'CO)$_2$O → PhN(R)COR' → HNO$_3$ → ortho-NO$_2$ + para-NO$_2$ products]

5. Aromatic diazonium salts, which are prepared by the diazotisation of primary aromatic amines, are key intermediates in the preparation of a range of aromatic compounds. In particular, the diazo group can be replaced by OH, Br, CN, Cl and I (nitrogen, N_2, is a good leaving group):

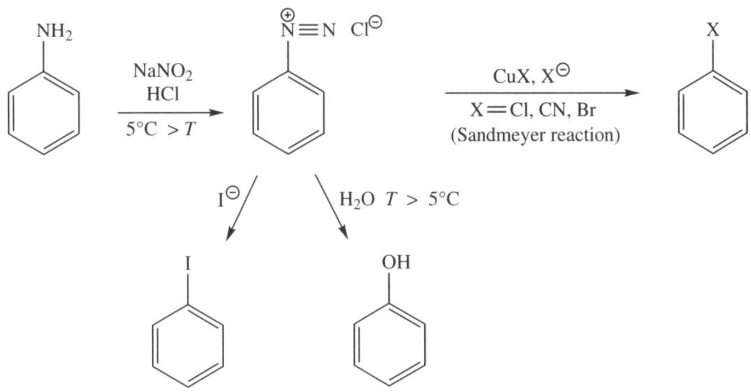

Problems

5.1 Why is aniline ($C_6H_5NH_2$) a much weaker base than ethylamine ($C_2H_5NH_2$)?

5.2 Describe briefly how you would prepare a solution of a diazonium salt from aniline ($C_6H_5NH_2$) and how this solution could be used to prepare the following compounds:

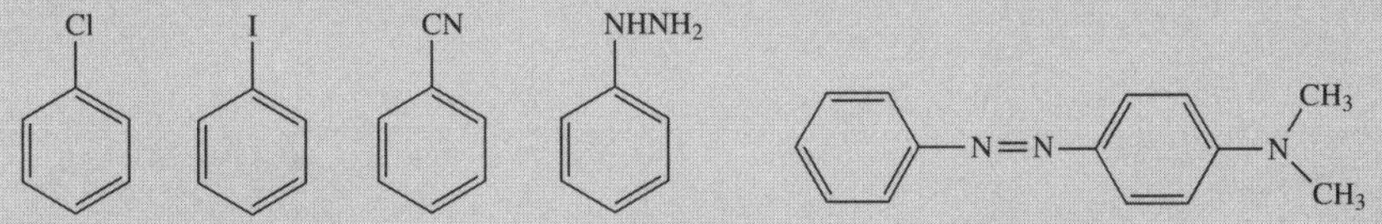

(Cl-C₆H₅, I-C₆H₅, CN-C₆H₅, NHNH₂-C₆H₅, C₆H₅-N=N-C₆H₄-N(CH₃)₂)

5.3 Give methods for the preparation of each of the following amines.

(a) 4-methylaniline (CH₃-C₆H₄-NH₂) (from toluene)

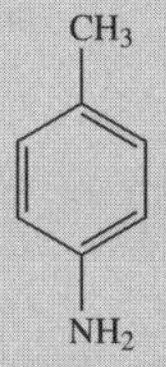

(b) $H_2NCH_2(CH_2)_4CH_2NH_2$ (from $ClCH_2(CH_2)_2CH_2Cl$)

(c) $C_6H_5\overset{H}{N}\!-\!CH(CH_3)_2$

5.4 Give reagents and conditions for each of the following conversions:

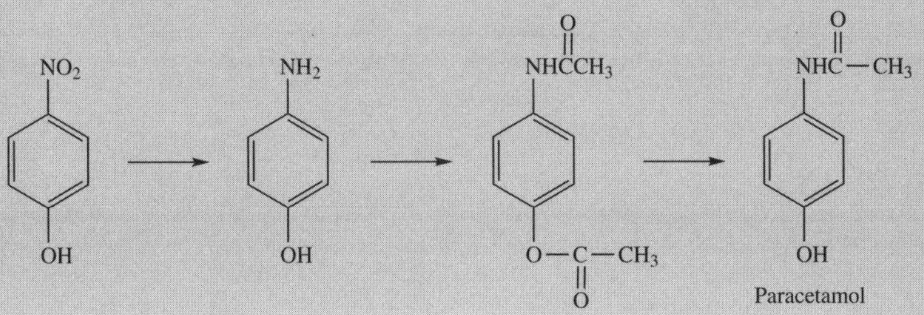

4-nitrophenol → 4-aminophenol → 4-acetamidophenyl acetate → Paracetamol (4-acetamidophenol)

5.5 Describe a simple chemical test, and the expected observations, that could distinguish between the following isomers:

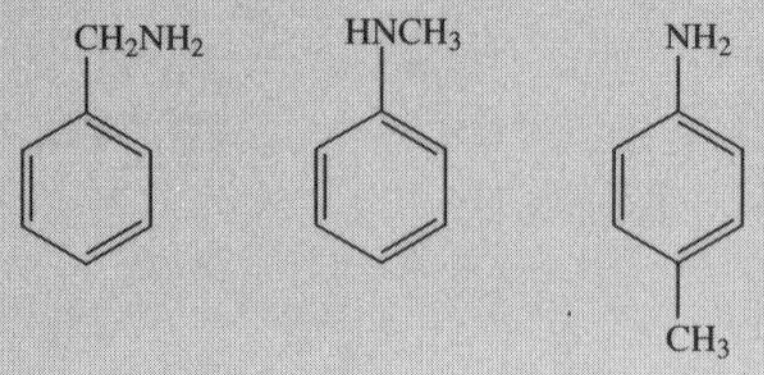

C₆H₅-CH₂NH₂   C₆H₅-HNCH₃   4-CH₃-C₆H₄-NH₂

6 Carbohydrates

Topics	
	6.1 D-Glucose
	6.2 D- and L-Sugars
	6.3 D-(−)-Fructose
	6.4 Reactions of sugars
	6.5 Oligosaccharides and polysaccharides
	6.6 Enzymic degradation of starch and cellulose
	6.7 Summary

The carbohydrates are a class of compounds with examples that occur in all living systems. The name carbohydrate was derived from the early observation that many members have molecular formulae of the type $C_m(H_2O)_n$, but the name is no longer applied in this restricted sense, and is used loosely to describe many aliphatic polyhydroxy compounds and their derivatives. The simple carbohydrates, or sugars, having the formula $C_nH_{2n}O_n$, ($n \geq 3$) are known as **monosaccharides**, and are polyhydroxyaldehydes or polyhydroxyketones; we will discuss their chemistry in relation to that of aldehydes and ketones (see Chapter 3). The groups of isomeric compounds $C_3H_6O_3$, $C_4H_8O_4$, $C_5H_{10}O_5$ and $C_6H_{12}O_6$ are known as **trioses**, **tetroses**, **pentoses** and **hexoses** respectively, and these groups comprise the majority of naturally occurring sugars. Examples of heptoses, octoses, etc., are known in nature but occur much less frequently.

In addition, a monosaccharide that contains an aldehyde group is known as an **aldose**, and one that contains a keto group is known as a **ketose**.

Monosaccharides cannot be hydrolysed to simpler compounds, whereas disaccharides, and polysaccharides, can be hydrolysed to two, or more, monosaccharide molecules.

6.1 D-Glucose

D-Glucose is one of the most abundant sugars in nature, and a study of its chemistry will serve to illustrate the chemistry of all the simple aldoses. Glucose is a monosaccharide, with the molecular formula $C_6H_{12}O_6$.

Glucose reacts with phenylhydrazine (to form a phenylhydrazone) and with hydrogen cyanide (to form a cyanohydrin), both of which are characteristic reactions of the carbonyl group:

$$C_6H_{12}O_6 \text{ Glucose} \xrightarrow{\text{PhNHNH}_2, \text{HOAc}} \underset{\text{Phenylhydrazone}}{R_2C=N-NHPh}$$

$$C_6H_{12}O_6 \text{ Glucose} \xrightarrow{\text{HCN}} \underset{\text{Cyanohydrin}}{R_2C(OH)(CN)}$$

Glucose is reduced by sodium borohydride to sorbitol, which forms a hexa-acetate (hexa-ester) on reaction with acetic anhydride. Glucose is oxidised under mild oxidation conditions (Br_2, H_2O) to the optically active gluconic acid, which reacts with hydrogen iodide (replacing all the hydroxy groups by hydrogens) to give *n*-hexanoic acid.

Since it is easily oxidised, glucose must be an aldehyde (aldose), which is unbranched since it is reduced to *n*-hexanoic acid.

Glucose is thus a pentahydroxy aldohexose and, since we know that two hydroxyl groups on the same carbon atom (*geminal*-diol) is an unstable arrangement compared with the corresponding carbonyl group, each hydroxy group must be on a different carbon atom:

$$R_2C(OH)_2 \rightleftharpoons R_2C=O + H_2O$$

Glucose, $C_6H_{12}O_6$, must therefore have the structure:

$$\underset{}{HC(=O)(CHOH)_4CH_2OH}$$

which has four asymmetric carbon atoms and so 2^4 (=16) stereoisomers, of which only one is glucose.

For his research leading to the structure of D-(+)-glucose, i.e. which of these 16 stereoisomers is glucose, Emil Fischer was awarded the Nobel prize in 1902.

```
   ¹CHO                    CHO
    |                       ┊
 H—²—OH                 H ──OH
    |                       ┊
HO—³—H                 HO ──H
    |            ≡         ┊
 H—⁴—OH                 H ──OH
    |                       ┊
 H—⁵—OH                 H ──OH
    |                       ┊
  ⁶CH₂OH                  CH₂OH

  D-(+)-glucose
```

Note that in the Fischer projection formula for glucose only the OH on carbon-3 is on the left-hand side.

The structure above is the open-chain form of glucose. We now realise that this structure does not adequately represent the properties of glucose. For example, glucose does not form a bisulphite addition compound, and reacts with methanol in the presence of anhydrous HCl to give two isomeric products, which contain only one methyl (CH_3) group, but resemble an acetal. This reaction is typical of a hemiacetal forming an acetal (see p. 89) and, in this case only one of the hydroxy groups undergoes reaction.

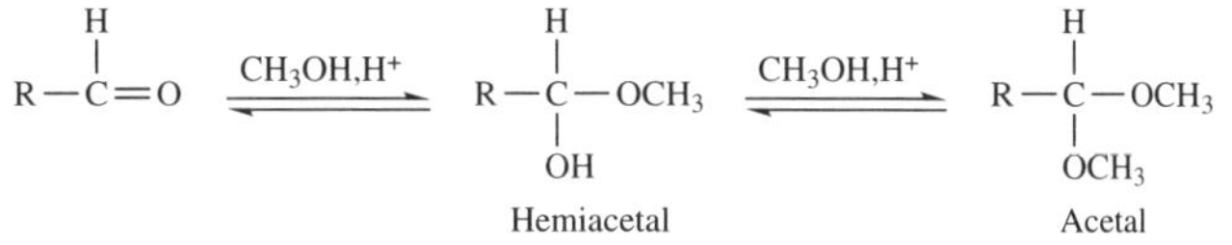

$$R-CHO \xrightarrow{CH_3OH, H^+} R-CH(OH)(OCH_3) \xrightarrow{CH_3OH, H^+} R-CH(OCH_3)_2$$

Hemiacetal → Acetal

Finally, glucose exists in two isomeric solid forms, known as **anomers**, which are obtained under different crystallization conditions:

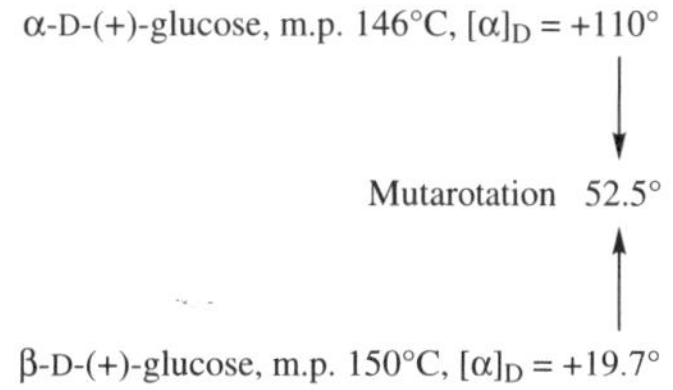

α-D-(+)-glucose, m.p. 146°C, $[\alpha]_D = +110°$

↓

Mutarotation 52.5°

↑

β-D-(+)-glucose, m.p. 150°C, $[\alpha]_D = +19.7°$

This difference in specific rotation is one of molecular structure and not of crystal structure, as in polymorphism. However, if freshly prepared solutions of α-D-(+)-glucose and β-D-(+)-glucose are allowed to stand, the rotatory powers of the solutions change slowly and eventually achieve a common value equivalent to a specific rotation of +52.5°. This phenomenon is known as **mutarotation**.

All of this evidence leads to the conclusion that the correct structures for hexoses (or pentoses) are cyclic ones, in which the carbonyl function is converted into a hemiacetal by combination with one of the hydroxyl groups in

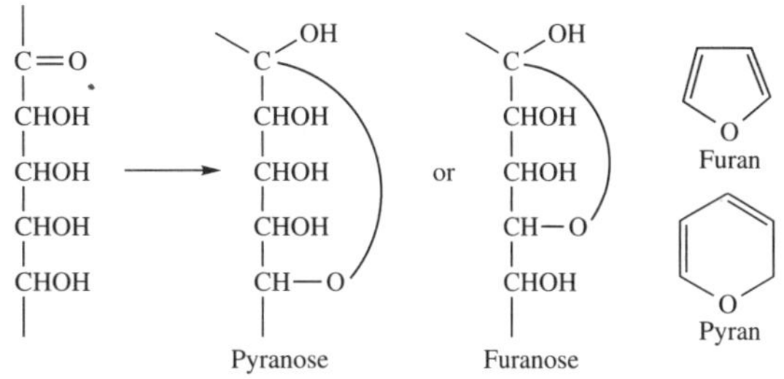

open chain → Pyranose or Furanose; Furan, Pyran

the same molecule. Normally only five- or six-membered rings are produced in this way, and are known as the furanose and pyranose structures respectively, after the parent heterocyclic compounds, furan and pyran. In this reaction, a new chiral centre is produced at the carbonyl carbon, so two diastereoisomeric (epimeric) hemiacetals are formed.

Glucose, in the free state, exists entirely as the pyranose form, formed by combination of the aldehyde (C-1) with the OH on C-5. The hydroxy group at C-1, formed in the cyclic hemiacetal formation, is the one different group (epimeric).

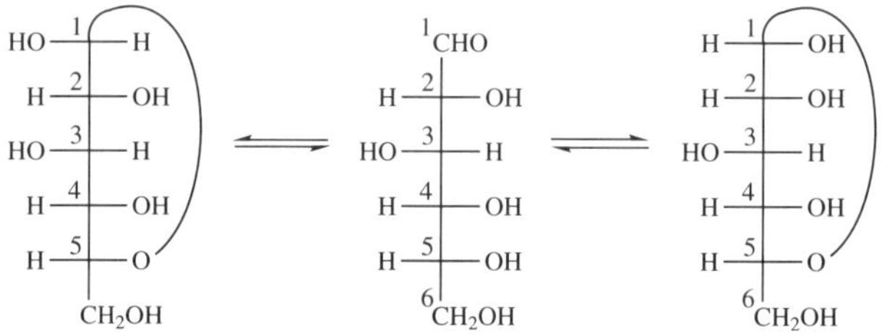

Mutarotation is caused by the slow equilibration of the C-1 epimers, probably via a very low concentration of the open-chain aldehyde form, which is responsible for the characteristic reactions of aldehydes.

Glucose, being a six-membered ring, exists primarily in the chair conformation with most of the substituents in the equatorial positions:

β-D-(+)-Glucose
[β-D-(+)-glucopyranose]

α-D-(+)-Glucose
[α-D-(+)-glucopyranose]

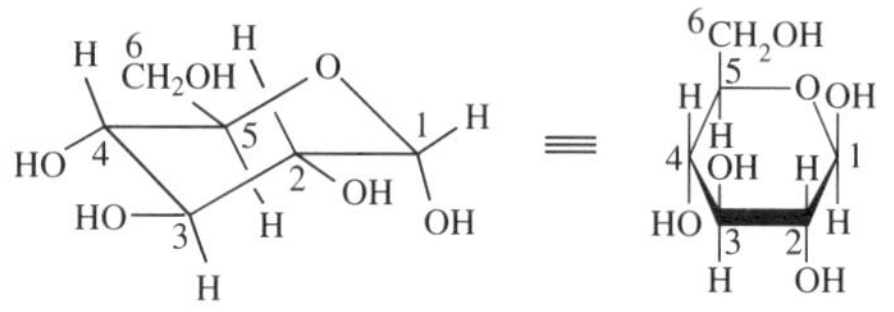

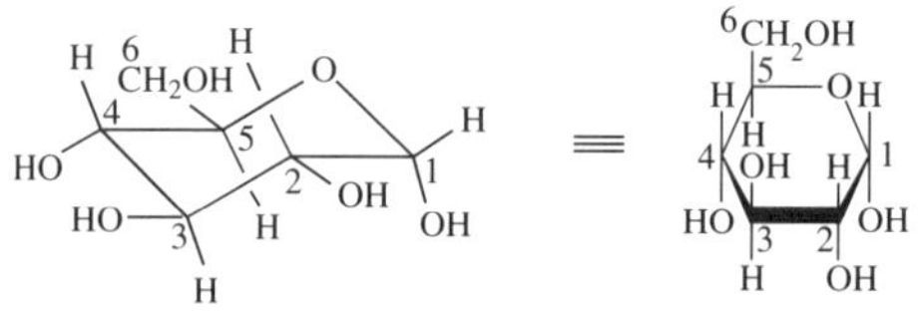

6.2 D- and L-Sugars

If we regard $C_3H_6O_3$ as the simplest molecular formula of a sugar, then there are three isomeric trioses, two of which are enantiomeric:

```
      CHO              CHO            CH₂OH
      |                |              |
H — C — OH       HO — C — H           CO
      |                |              |
      CH₂OH            CH₂OH          CH₂OH

D-Glyceraldehyde   L-Glyceraldehyde   Dihydroxyacetone
R-Glyceraldehyde   S-Glyceraldehyde   (dihydroxypropanone)
```

We have already come across the R,S nomenclature but the much-used D,L nomenclature now needs explaining further. When the Fischer projection of a sugar is drawn with the carbon chain vertical and the aldehyde or ketone function at the top, D-sugars (by definition) have the hydroxyl group on the RHS of the chain, while L-sugars have this hydroxyl on the LHS of the chain.

```
    ¹CHO           ¹CHO         ¹CH₂OH        ¹CH₂OH
     |              |            |             |
   (CHOH)ₙ        (CHOH)ₙ       ²C=O          ²C=O
     |              |            |             |
 H—|ⁿ⁺²OH    HO—|ⁿ⁺²H         (CHOH)ₙ       (CHOH)ₙ
     |              |            |             |
    CH₂OH         CH₂OH      H—|ⁿ⁺³OH     HO—|ⁿ⁺³H
                                 |             |
                                CH₂OH         CH₂OH

   D-aldoses      L-aldoses    D-ketoses     L-ketoses
```

Thus, D- and L-glyceraldehyde may be regarded as the parents of two series of aldose sugars, and dihydroxyacetone as the parent of the ketose sugars. It should be noted that the specific names given to the stereoisomeric sugars imply a sequence of chiral centres of fixed relative configuration. Thus, D-ribose and L-ribose do not differ merely in the configuration of a single chiral centre, but have mirror-image related structures, in which the configurations of all the chiral centres have changed. The epimer of D-ribose in which only the configuration of the penultimate carbon is changed is L-lyxose (cf. the Fischer projection of D-lyxose).

```
      CHO              CHO              CHO
      |                |                |
H — C — OH       HO — C — H       H — C — OH
      |                |                |
H — C — OH       HO — C — H       H — C — OH
      |                |                |
H — C — OH       HO — C — H       HO — C — H
      |                |                |
      CH₂OH            CH₂OH            CH₂OH

    D-Ribose          L-Ribose         L-Lyxose
```

142   Carbohydrates

D-*Ketoses*

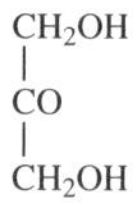

$$\begin{array}{c} CH_2OH \\ | \\ CO \\ | \\ CH_2OH \end{array}$$
Dihydroxyacetone

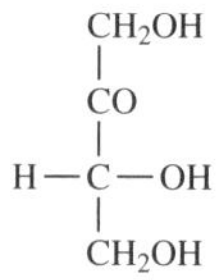

$$\begin{array}{c} CH_2OH \\ | \\ CO \\ | \\ H-C-OH \\ | \\ CH_2OH \end{array}$$
D-Erythrulose

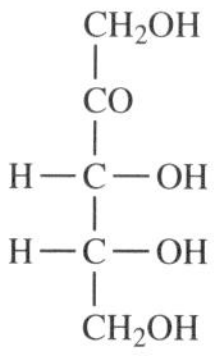

D-Ribulose 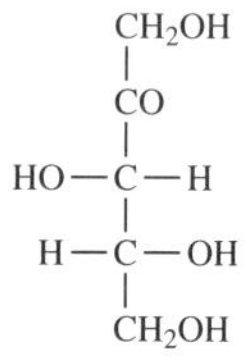   D-Xylulose

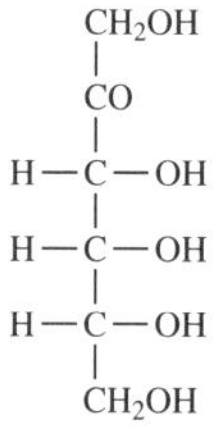

D-Allulose   D-Fructose 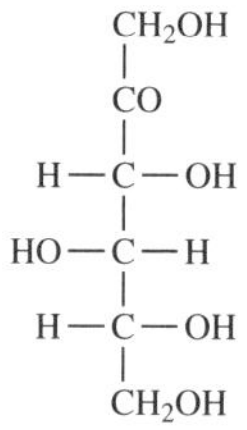   D-Sorbose 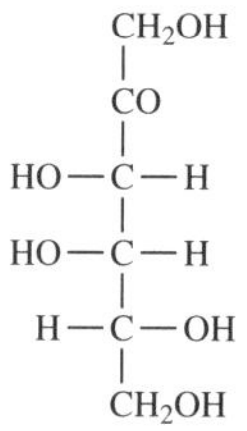   D-Tagatose
(D-Psicose)

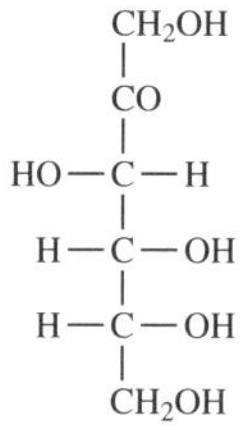

(See D-Aldoses on facing page)

6.3   D-(−)-Fructose

The most important ketose is D-(−)-fructose on account of its occurrence in sucrose (p.151). Like glucose, fructose exists predominantly in the form of a

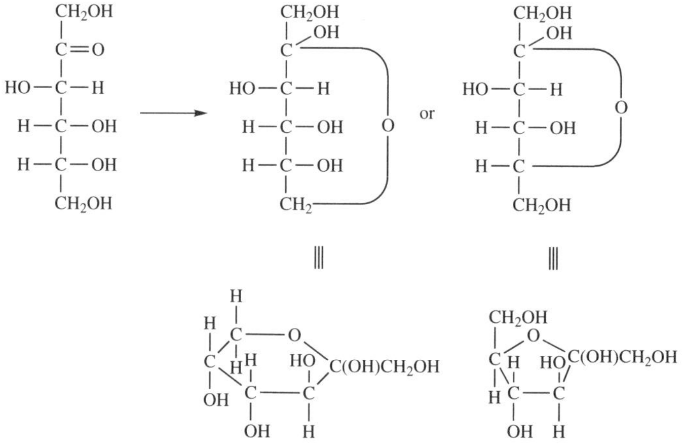

D-Fructopyranose   D-Fructofuranose

D-*Aldoses*

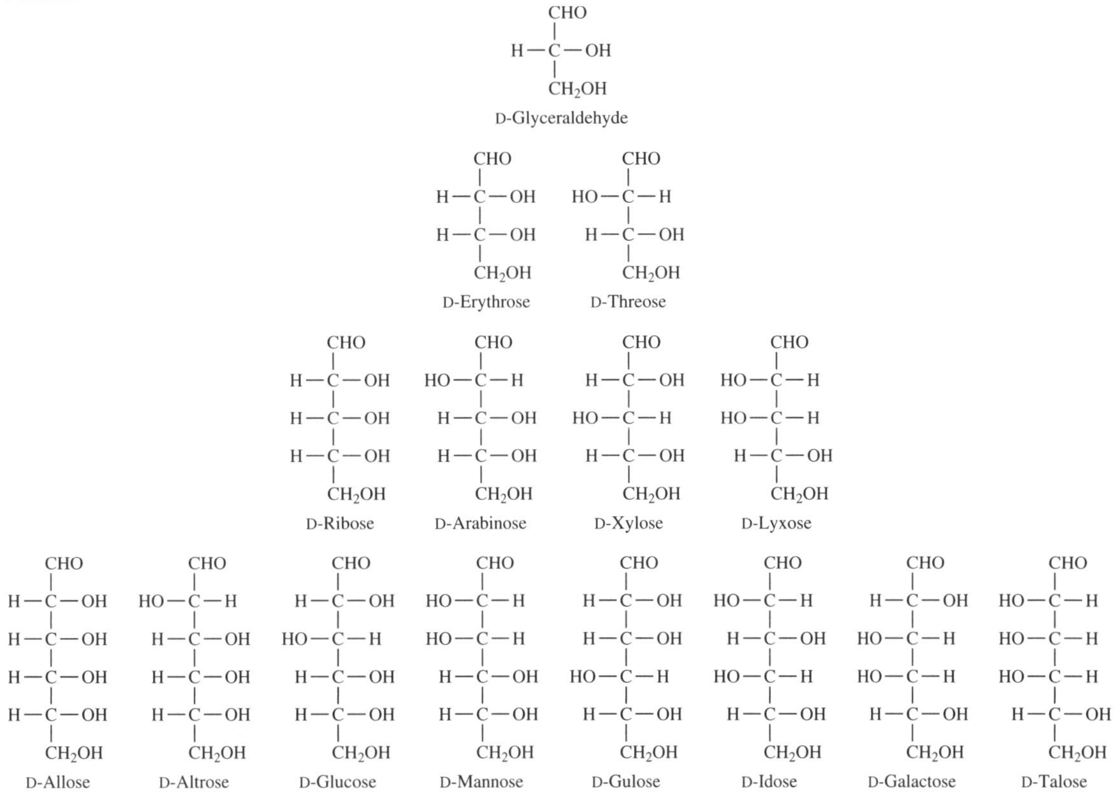

$$
\begin{array}{c}
\text{CHO} \\
| \\
\text{H}-\text{C}-\text{OH} \\
| \\
\text{CH}_2\text{OH}
\end{array}
$$
D-Glyceraldehyde

D-Erythrose   D-Threose

D-Ribose   D-Arabinose   D-Xylose   D-Lyxose

D-Allose   D-Altrose   D-Glucose   D-Mannose   D-Gulose   D-Idose   D-Galactose   D-Talose

cyclic hemiacetal and exhibits mutarotation. Crystalline fructose is entirely in the pyranose form, but derivatives of both pyranose and furanose isomers are known.

6.4 Reactions of sugars

6.4.1 Cyanohydrin formation

One of the few addition and condensation reactions (characteristic of aldehydes) shown by glucose is the addition of hydrogen cyanide to form epimeric cyanohydrins:

$$\underset{R}{\overset{R}{>}}\!\!=\!\!O \xrightarrow{\text{HCN}} \underset{R}{\overset{R}{>}}\!\!\!<\!\!\!\overset{\text{OH}}{\underset{\text{CN}}{}}$$

For sugars this reaction is especially important since it is a key step in the Killiani–Fischer synthesis – a method of lengthening the chain of aldose sugars.

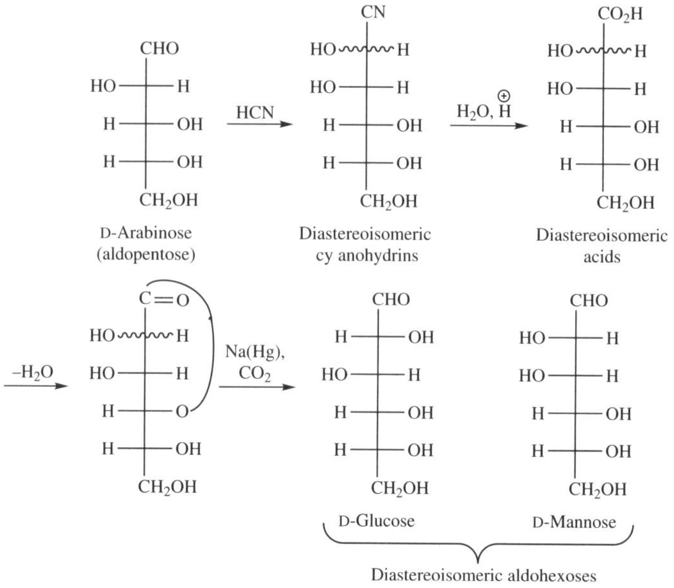

The cyanohydrins (hydroxynitriles) formed by the addition of HCN are hydrolysed to the epimeric acids and these are relatively easy to separate. The acids cyclise to the five-membered lactones (esters) which can be reduced to the epimeric aldohexoses.

6.4.2 Osazone formation

Glucose reacts with phenylhydrazine to give the phenylhydrazone, which is too soluble to be isolated. A further reaction with two molecules of phenylhydrazine then occurs, in which one molecule of phenylhydrazine is reduced to phenylamine (aniline) and ammonia and the phenylhydrazone is oxidised to the bis-phenylhydrazone derived from the α-dicarbonyl compound. These compounds, known as osazones, were extensively used in the past to characterise and identify sugars.

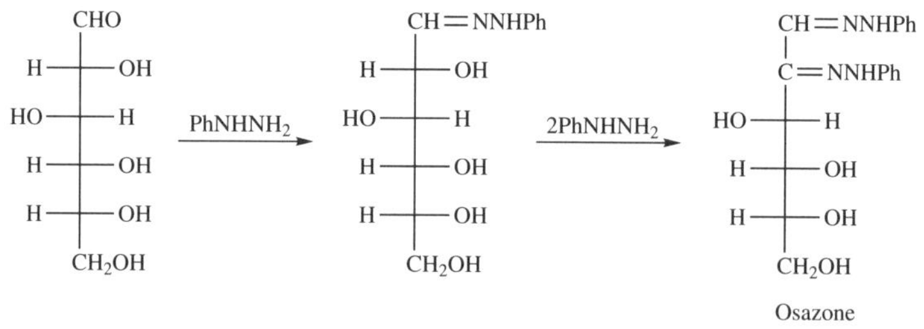

Note that, in the formation of an osazone, one of the chiral centres is destroyed so that aldoses epimeric at C-2, e.g. D-glucose and D-mannose, form the same osazone, as does D-fructose.

6.4.3 Oxidation

The oxidation of glucose with dilute nitric acid or bromine water gives, initially, D-gluconic acid. Further oxidation, for example in nitric acid, gives D-gluco-saccharic acid.

```
      CHO                       CO₂H                       CO₂H
   H──OH                     H──OH                      H──OH
  HO──H      Br₂/H₂O       HO──H        HNO₃         HO──H
   H──OH     ────────→      H──OH       ────────→     H──OH
   H──OH                     H──OH                     H──OH
     CH₂OH                     CH₂OH                     CO₂H
   D-Glucose              D-Gluconic acid          D-Gluco-saccharic acid
                          (an aldonic acid)          (an aldaric acid)
```

Oxidation to the gluconic acid is a key step in the Ruff degradation, a means of shortening the carbon chain of aldoses by one carbon atom.

```
      CHO                     CO₂H
   H──OH                   H──OH                             CHO
  HO──H     Br₂/H₂O      HO──H       1. CaCO₃            HO──H
   H──OH    ────────→     H──OH      2. H₂O₂/Fe³⁺         H──OH
   H──OH                   H──OH     ──────────→          H──OH
     CH₂OH                   CH₂OH                          CH₂OH
   D-Glucose              Aldonic acid                   D-Arabinose
  (aldohexose)                                           (aldopentose)
```

Oxidation is also easily achieved using Fehling's or Benedict's solutions, and ammoniacal silver nitrate.

6.4.4 Reduction

Aldoses and ketoses are readily reduced (H_2/Ni, $NaBH_4$ or Na(Hg)/H_2O) to the corresponding alcohols:

```
      CHO                     CH₂OH                    CH₃
   H──OH                   H──OH                    H──H
  HO──H        redn      HO──H       HI/heat        H──H
   H──OH     ────────→    H──OH      ────────→      H──H
   H──OH                   H──OH                    H──H
     CH₂OH                   CH₂OH                    CH₃
                       Glucitol (sorbitol)
```

Further reduction, using HI/heat, can be accomplished to give the corresponding alkane.

6.4.5 Reaction with alcohols

As mentioned previously, glucose reacts with alcohols, in the presence of acid, to form two isomeric products. Being a cyclic hemiacetal, this converts glucose into the acetals, or glycosides. The isomers occur as a result of the anomers at the C-1 position:

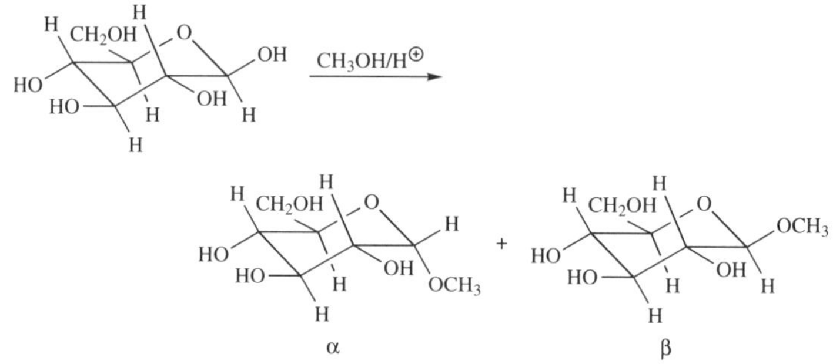

Methyl-D-glucopyranosides
(methyl-D-glucosides)

6.4.6 Miscellaneous reactions

The reaction of glucose with acylating agents, e.g. acid anhydrides, gives two anomeric penta-acyl derivatives. In these derivatives the acyl group attached to C-1 is very readily removed by hydrolysis since it is the acyl derivative of a hemiacetal, and the other acylated groups show the normal reactivity of esters.

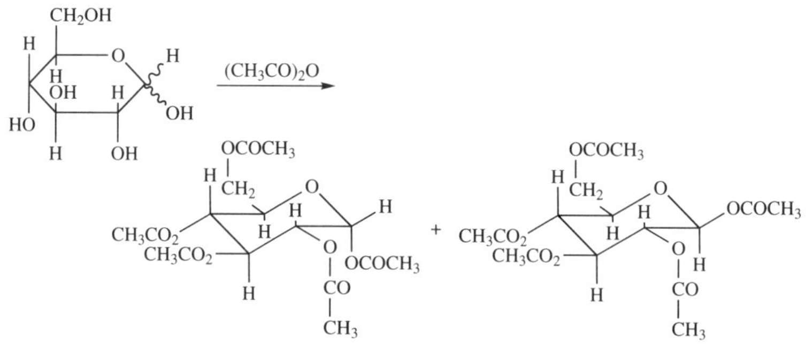

The reaction of glucose with thiols results in the ring-opening of the pyranose ring and the formation of a normal thioacetal:

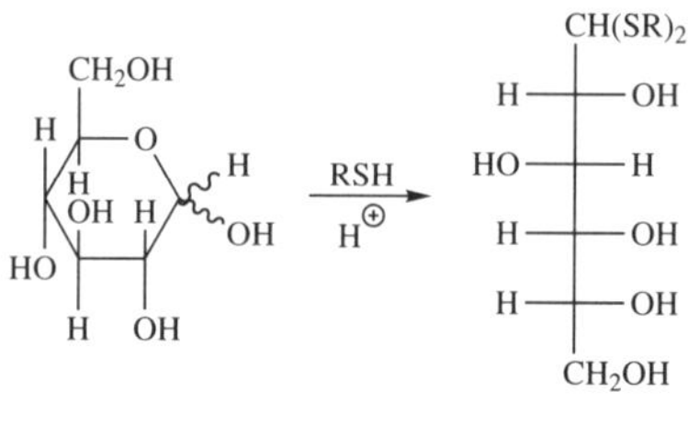

6.4.7 Reaction with alkali

The action of strong alkalis on glucose leads to brown resinous products, but in weakly alkaline solutions, D-glucose rearranges to form a mixture of hexoses, in which D-glucose, D-mannose and D-fructose predominate. It is very likely that enolisation is responsible for the interconversion, which must occur by way of the open-chain aldehyde.

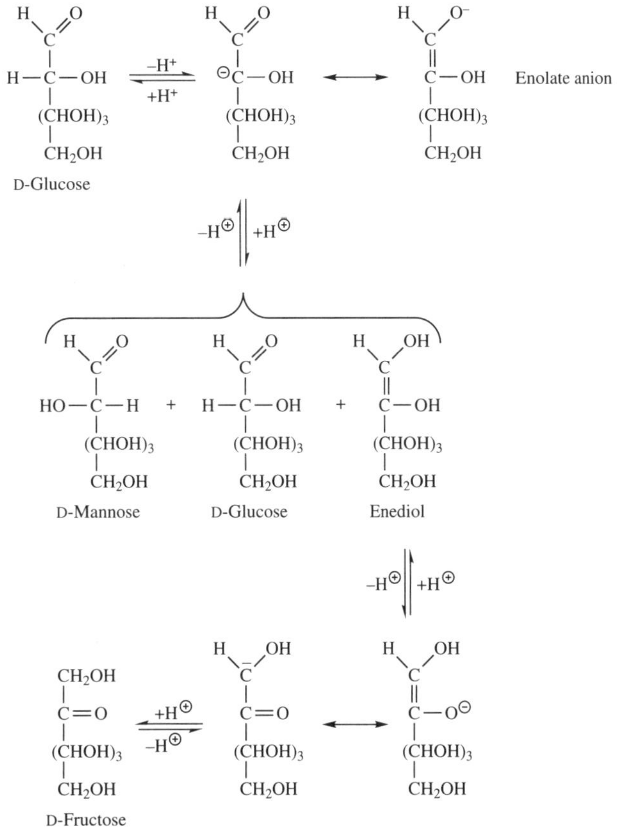

[Scheme showing D-Glucose ⇌ enol ⇌ Enolate anion, interconverting via enediol to give D-Mannose + D-Glucose + Enediol, and further to D-Fructose via its enol and enolate.]

Removal of a proton from C-2 of D-glucose forms an enolate anion which can be reprotonated on C-2 to form either D-glucose or D-mannose, depending upon the direction from which the incoming proton approaches the enolate:

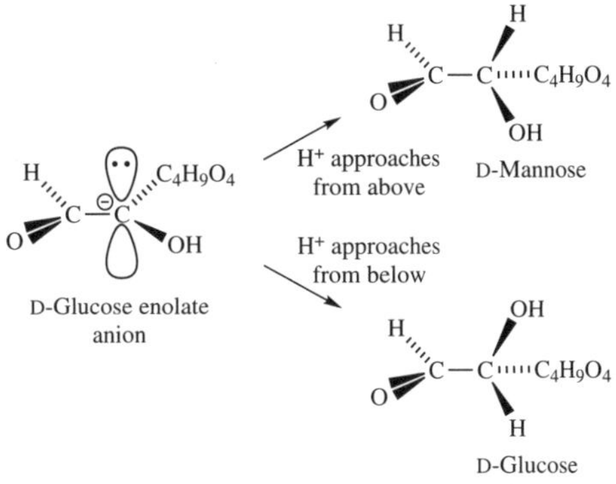

[Diagram: D-Glucose enolate anion; H$^+$ approaches from above → D-Mannose; H$^+$ approaches from below → D-Glucose, with $C_4H_9O_4$ substituent.]

Alternatively, the enolate anion derived from glucose may be protonated on oxygen to form an enediol, which, by loss of a proton from the hydroxyl group on C-2, can be converted into a new enolate anion derived from D-fructose. The enediol formed as the intermediate in this process is the enol form of D-glucose, D-mannose and D-fructose, and the interconversion of these sugars is just a special case of the process responsible for the racemisation of enantiomers with a chiral centre adjacent to a carbonyl group. A similar sequence of reactions will convert D-fructose into its epimer D-allulose.

Reactions of this type are known to occur in biochemical processes, and two examples taken from the photosynthetic cycle are given below. It should be noted that in both cases the hydroxymethylene group which undergoes the change is adjacent to a carbonyl group.

$$
\begin{array}{ccc}
\text{CH}_2\text{OH} & \text{CH}_2\text{OH} & \text{CHO} \\
| & | & | \\
\text{CO} & \text{CO} & \text{H—C—OH} \\
| & \xrightarrow{\text{Phosphoketo-epimerase}} | \xleftarrow{\text{Phosphopentose isomerase}} | \\
\text{HO—C—H} & \text{H—C—OH} & \text{H—C—OH} \\
| & | & | \\
\text{H—C—OH} & \text{H—C—OH} & \text{H—C—OH} \\
| & | & | \\
\text{CH}_2\text{OPO}_3\text{H}_2 & \text{CH}_2\text{OPO}_3\text{H}_2 & \text{CH}_2\text{OPO}_3\text{H}_2 \\
\text{D-Xylulose-5-phosphate} & \text{D-Ribulose-5-phosphate} & \text{D-Ribose-5-phosphate}
\end{array}
$$

Another reaction observed under alkaline conditions is the degradation of monosaccharides into smaller units by the reverse of the aldol reaction. Typical of this reaction is the cleavage of D-fructose into dihydroxyacetone and D-glyceraldehyde:

$$
\begin{array}{c}
\text{CH}_2\text{OH} \\
| \\
\text{C=O} \\
| \\
\text{HO—C—H} \\
| \\
\text{H—C—O—H} \quad \text{Base} \\
| \\
\text{H—C—OH} \\
| \\
\text{CH}_2\text{OH}
\end{array}
\longrightarrow
\begin{array}{c}
\text{CH}_2\text{OH} \\
| \\
\text{C—O}^{\ominus} \\
\| \\
\text{H—C—OH}
\end{array}
+
\begin{array}{c}
\text{H—C=O} \\
| \\
\text{H—C—OH} \\
| \\
\text{CH}_2\text{OH}
\end{array}
$$

$$\downarrow +\text{H}^+$$

$$
\begin{array}{c}
\text{CH}_2\text{OH} \\
| \\
\text{C=O} \\
| \\
\text{CH}_2\text{OH}
\end{array}
$$

Compare this mechanism with that of the aldol reaction (p. 93). The enzymic cleavage of fructose-1,6-bisphosphate into D-glyceraldehyde-3-phosphate and dihydroxyacetone phosphate is an important step in anaerobic glycolysis.

6.5 Oligosaccharides and polysaccharides

The carbohydrates include many substances more complex than the simple sugars, and numerous naturally occurring carbohydrates are built up by combination of two or more sugar molecules. The names of the various classes of carbohydrate indicate the number of simple sugar (monosaccharide) molecules from which the carbohydrate is constructed (e.g. disaccharides and trisaccharides) while the terms oligosaccharide (oligo = few) and polysaccharide (poly = many) are used to denote compounds derived from small and large numbers of monosaccharide units respectively. Although many common polysaccharides are built up from hexoses, polysaccharides containing tetroses and pentoses are also well known.

The formation of polysaccharides from monosaccharides can be formally regarded as the conversion of the hemiacetal function of a simple sugar into an acetal (a glycoside) by combination with one of the hydroxyl groups of another sugar molecule. The result is a sequence of sugar residues linked by oxygen atoms. Just as the conversion of glucose into methyl glucoside results in two anomeric products (p. 146), so the stereochemistry of the oxygen bridge between sugar residues can differ, and profound differences in biological properties exist between polysaccharides that differ only in the stereochemistry of the ether bridge.

Some of the better known disaccharides are described below.

6.5.1 Maltose

Maltose, a disaccharide produced by the enzymic degradation of starch, is formed by the combination of two molecules of D-glucose. Both glucose residues have pyranose rings, and the two residues are linked by combination of the hemiacetal group of one with the hydroxyl group on C-4 of the second. Note that the stereochemistry of the link is that of an α-glucoside (p. 146).

Since the hemiacetal function of the second (right-hand) glucose moiety is not involved in bonding, maltose is a reducing sugar (i.e. it reduces Fehling's solution or ammoniacal silver nitrate), and forms α and β anomers.

6.5.2 Cellobiose

Cellobiose, a disaccharide obtained by the chemical degradation of cellulose, is constructed from two molecules of D-glucose, linked in a manner very similar to that in maltose, but differing in the stereochemistry of the oxygen bridge, cellobiose being a β-glucoside. Like maltose, cellobiose has a free hemiacetal function, and is therefore a reducing sugar.

The drawing of full constitutional formulae for disaccharides or more complex oligosaccharides is tedious. For many purposes the following shorthand method of representing chains of sugar residues has been found convenient:

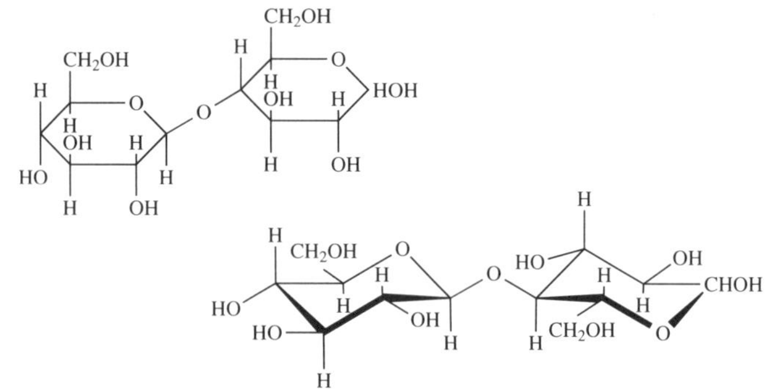

$$\text{glucose} \xrightarrow{1\alpha \quad 4} \text{glucose} \quad \text{i.e. maltose}$$

$$\text{glucose} \xrightarrow{1\beta \quad 4} \text{glucose} \quad \text{i.e. cellobiose}$$

The symbols over the joining line indicate the position and stereochemistry of the glycosidic link.

The biological importance of the stereochemistry of the links between sugar residues in polysaccharides can be judged by a comparison between maltose and cellobiose. Maltose is hydrolysed by maltase, an enzyme present in yeast, which can also hydrolyse many derivatives of α-D-glucose, e.g. α-methyl-D-glucoside. However, maltase is quite unable to hydrolyse cellobiose, even though the molecules are so similar. Cellobiose is hydrolysed to glucose by emulsin, an enzyme occurring in almonds, which is specific for β-glucosides but has no effect on maltose or other α-glucosides. The reason for this specificity is very probably that, prior to reaction, the enzyme adsorbs the substrate at a special region on the surface of the protein molecule (the 'active site') and this combination of enzyme and substrate is critically dependent on the three-dimensional structure. Alteration of the stereochemistry of the substrate produces a molecule that does not fit the active site of the enzyme and is therefore not susceptible to attack. The biological resolution of racemic mixtures (p. 22) depends upon this stereospecificity.

6.5.3 Gentiobiose (glucose $\xrightarrow{1\beta \quad 6}$ glucose)

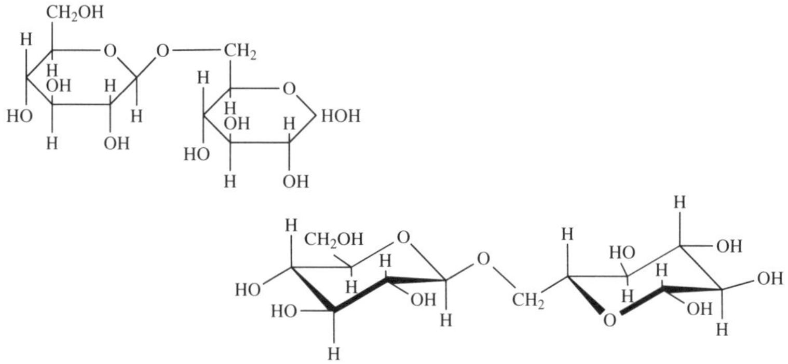

Gentiobiose, also formed from two molecules of D-glucose, is an example of a disaccharide in which the hydroxyl group on C-6 of a glucose unit is involved in the glycoside link. Gentiobiose is a reducing sugar, and, since it is a β-glucoside, is hydrolysed by emulsin.

6.5.4 Lactose (galactose $\xrightarrow{1\beta\ 4}$ glucose)

Lactose is a disaccharide, present in milk, formed from one molecule of D-galactose and one of D-glucose, joined by a β linkage. It is not hydrolysed by either emulsin or maltase, since it is not a D-glucoside, but can be cleaved by β-galactosidases, which are specific for derivatives of β-D-galactose.

6.5.5 Sucrose (glucose $\xrightarrow{1\alpha\ 2\beta}$ fructose)

Sucrose (cane sugar) is commercially the most important of the simple carbohydrates, and is extracted from the juice of sugar cane or sugar beet. It is formed by combination of one molecule of D-glucose with one molecule of D-fructose, but unlike previous examples, both hemiacetal groups are involved in this linkage. Since acetals and glycosides are stable to alkali and no aldehyde or ketone function (or its equivalent) is present in the molecule, sucrose is a non-reducing sugar, being unaffected by Fehling's solution or ammoniacal silver nitrate. However, boiling for a few minutes with dilute mineral acid cleaves the glycosidic linkage, liberating free glucose and fructose, which will then readily reduce these reagents.

Since the glycosidic linkage involves both hemiacetal functions, four stereoisomers are possible:

$$\text{glucose} \xrightarrow{1\alpha\ 2\alpha} \text{fructose}$$
$$\text{glucose} \xrightarrow{1\beta\ 2\alpha} \text{fructose}$$
$$\text{glucose} \xrightarrow{1\alpha\ 2\beta} \text{fructose}$$
$$\text{glucose} \xrightarrow{1\beta\ 2\beta} \text{fructose}$$

Hydrolysis of sucrose by enzymes of known stereospecificity shows that sucrose is an α-glucoside and a β-fructoside.

6.5.6 Cellulose

Cellulose $(C_6H_{10}O_5)_n$ is the principal polysaccharide of the cell walls of higher plants, and the rigidity of plant tissue is due to its presence. Cellulose consists of very long chains of D-glucose residues linked (1β–4) with the resultant molecule having a relative molecular mass of 10^5–10^6 depending upon the plant source.

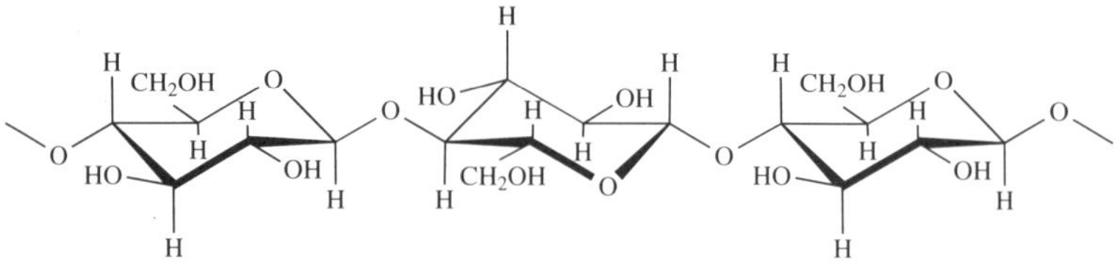

Part of a cellulose molecule

Cellulose fibres consist of bundles of polysaccharide chains held together by hydrogen bonding. This arrangement gives the fibres their high mechanical strength.

6.5.7 Starch

Starch is the reserve polysaccharide of plants, being deposited in characteristic granules. These starch granules contain two components, amylose and amylopectin, both of which are built entirely of D-glucose residues.

Amylose, the minor constituent of starch, is the polysaccharide that gives the characteristic blue colour with iodine. It consists of a long chain of about 300 glucose residues linked (1α–4) as in maltose (p. 149). This long chain is thought to be coiled up into a helix in which there are about six glucose residues per turn (i.e. a helix of approximately 50 turns). The blue iodine complex is considered to be formed by insertion of iodine atoms or molecules into the axial cavity of the helix.

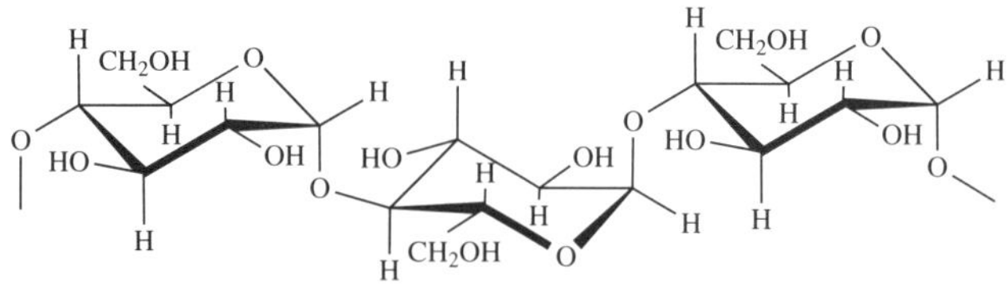

Part of an amylose molecule

Amylopectin is the chief constituent of most starches, and is a branched chain polysaccharide. Short chains of approximately 25 D-glucose residues linked (1α–4) are joined by (1α–6) links.

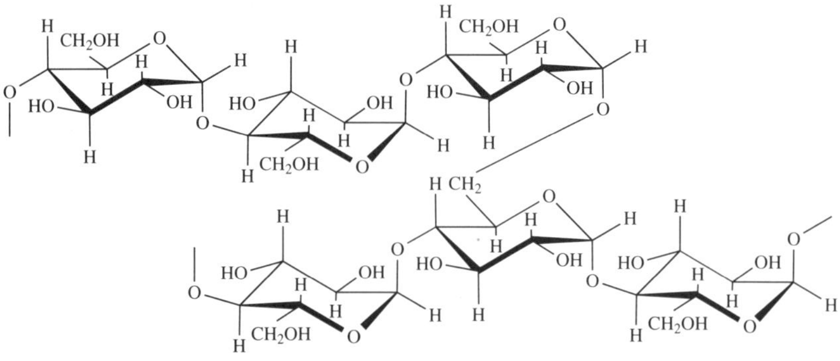

Part of an amylopectin molecule

This type of structure can best be represented by Figure 6.1, where the horizontal lines represent chains of (1α–4) linked glucose units and the vertical sections represent (1α–6) links between adjacent chains.

Figure 6.1

6.5.8 Glycogen

Glycogen (liver starch) is the principal reserve polysaccharide of animals, being deposited in the liver and muscles. Its structure is similar to that of amylopectin, but it has shorter chains of approximately 12 (1α–4)-linked glucose residues with (1α–6) cross-links.

The storage of glucose by conversion into these insoluble, polymeric forms avoids the problems of high osmotic pressure which would arise from the storage of comparable quantities of highly soluble simple sugars.

6.6 Enzymic degradation of starch and cellulose

Cellulose, amylose, amylopectin and glycogen are all polysaccharides constructed solely from D-glucose units, differing only in the position and stereochemistry of the glycosidic links. Comparison of the enzymic degradation of these compounds provides an indication of the remarkable specificity of enzymes.

Cellulose can be degraded by a group of enzymes known as 'cellulases', specific for (1β–4)-linked D-glucose polymers. Very few of the higher animals secrete cellulases in the digestive tract, and herbivores, for which cellulose is a major constituent of the diet, rely on symbiotic, cellulase-releasing, microorganisms to degrade cellulose in the food.

A group of enzymes known as 'amylases' are responsible for the hydrolysis of starch, and these are mostly specific for α-linked D-glucose polymers but have no action on cellulose. Several types of amylase are known, with varying substrate specificity. Exo- and endo-amylase specifically catalyse the hydrolysis of (1α–4)-linked D-glucose chains to maltose, and differ in the position of attack. Exo-amylase degrades the chain starting from a free end, whereas endo-amylase is able to attack in the middle of a chain. Either of these enzymes will degrade amylose extensively, but complete degradation requires the presence of yet another enzyme, 'Z-enzyme', which is known to be specific for certain types of β-glucoside, indicating the presence of a small number of β-links in the amylose molecule.

If exo-amylase acts on glycogen or amylopectin, the free ends of the (1α–4)-linked glucose chains are progressively degraded until a (1–6) link is encountered, when exo-amylase is no longer effective. Thus exo-amylase degrades the starch molecule into a simpler, but still large, molecule known as a 'limit dextrin'. Endo-amylase on the other hand, not requiring a free end for attack, can hydrolyse amylopectin and glycogen almost completely, giving predominantly maltose and isomaltose (glucose$\underline{1\alpha-6}$glucose). Thus, complete enzymic degradation of starch into glucose requires exo-amylase and endo-amylase, which degrade the polymer into disaccharides, and Z-enzyme, maltase and amylo-1,6-glucosidase, which acts on isomaltose, to convert the disaccharides into glucose.

Numerous other polysaccharides are known in nature, consisting of chains of condensed sugar units of varying types. Araban (poly-L-arabinose), found in association with pectin, and xylan (poly-D-xylose), found in woody plant tissue, are examples of polypentoses. Starch and cellulose are examples of polyhexoses, but many other types are known. Some micro-organisms produce dextrans ((1α–6)-linked poly-D-glucopyranose), and mannans (D-mannose chains) are found in the wood of some conifers. Galactans (poly-D-galactose) are also known. Inulin is a (2β–1)-linked poly-D-fructofuranoside found in the tubers of dahlias and other plants.

Besides polymers of simple carbohydrates, chains of modified sugars are known to occur widely. Pectins, which are constituents of the cell wall of plants, consist of chains of D-glucuronic acid partly in the form of its methyl ester; and chitin, a polysaccharide found in the shells of lobsters, crabs and

D-Glucuronic acid

N-Acetylglucosamine

cockroaches, is derived from N-acetylglucosamine. The cell walls of Gram-positive micro-organisms contain large amounts of a muropeptide, the structure of which consists of chains of alternate (1β–4)-linked residues of N-acetylglucosamine and N-acetylmuramic acid cross-linked by peptide chains:

N-Acetylmuramic acid (derived from N-acetylglucosamine and L-lactic acid)

6.7 Summary

1. Monosaccharides are polyhydroxyaldehydes or ketones but, although they show some of the characteristic reactions of the carbonyl group, they should be regarded as cyclic hemiacetals.

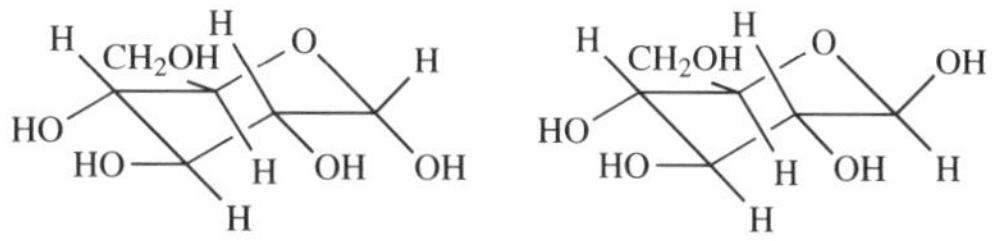

α-D-(+)-glucose β-D-(+)-glucose

Polysaccharides are composed of linked monosaccharide units and can be degraded to these units.

2. Sugars are classed in optical families, D- or L-, depending upon whether they are formally derived from D- or L-glyceraldehyde, i.e. dependent upon the position of the OH on the penultimate carbon of the carbon chain (drawn vertically in the Fischer projection with C-1 at the top).

D-glyceraldehyde (OH on RHS)   L-glyceraldehyde (OH on LHS)

3. Among the reactions undergone by sugars which are characteristic of the carbonyl group are: cyanohydrin formation (important in the Killiani–Fischer synthesis for lengthening the carbon chain), oxidation (important in the Ruff degradation), the reaction with phenylhydrazine (to form osazones), reduction and the reaction with alcohols (to form acetals-glycosides).

4. Oligo- and polysaccharides comprise monosaccharide units linked through C—O bonds:

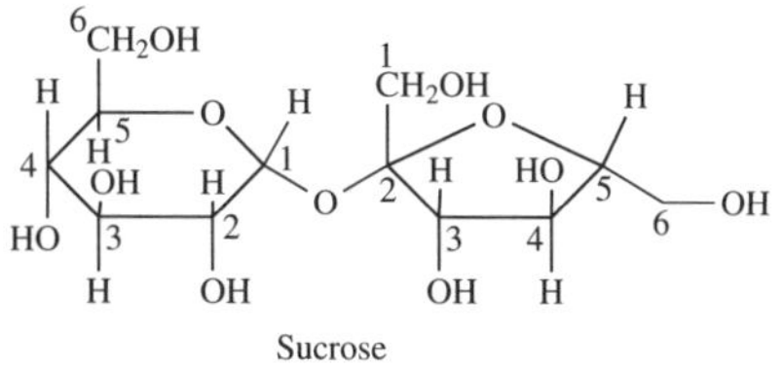

Sucrose
(glucose $\xrightarrow{1\alpha \quad 2\beta}$ fructose)

156   Carbohydrates

Problems

6.1 Draw the products of the reactions of D-glucose and D-fructose with the following reagents:

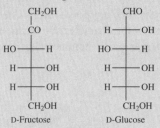

(a) 3 PhNHNH$_2$
(b) NaBH$_4$ (how are the two products formed by the reaction with D-fructose related?)
(c) MeOH, H$^+$
(d) NaCN, H$_2$SO$_4$

6.2 Draw a cyclic form of both D-fructose and D-glucose.

6.3 Give the shorthand nomenclature for the following disaccharides:

(a) cellobiose

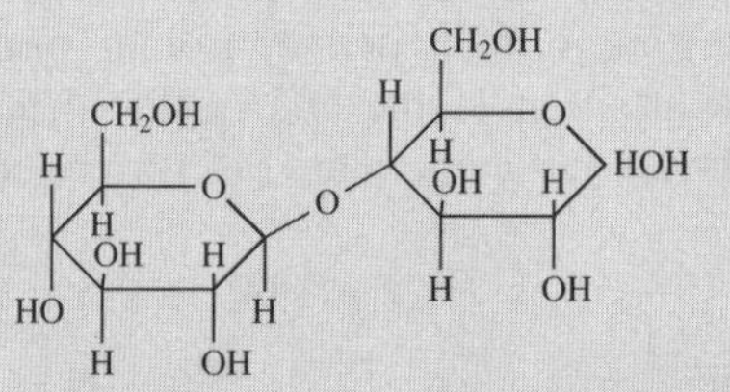

(b) maltose

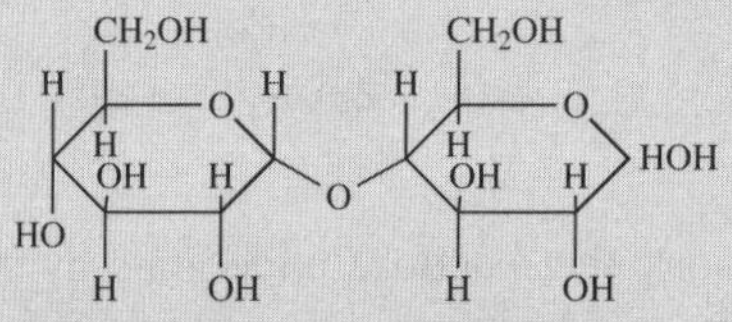

(c) sucrose

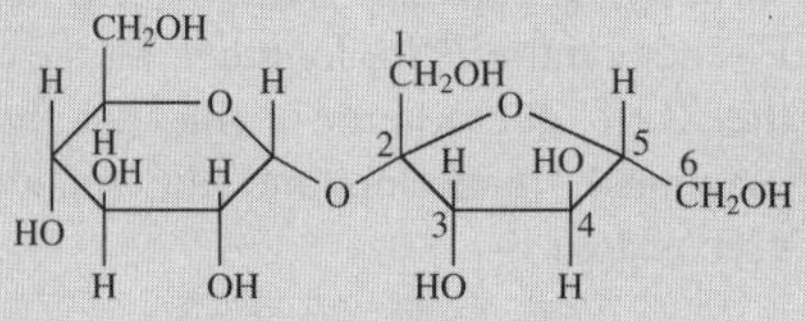

(d) lactose

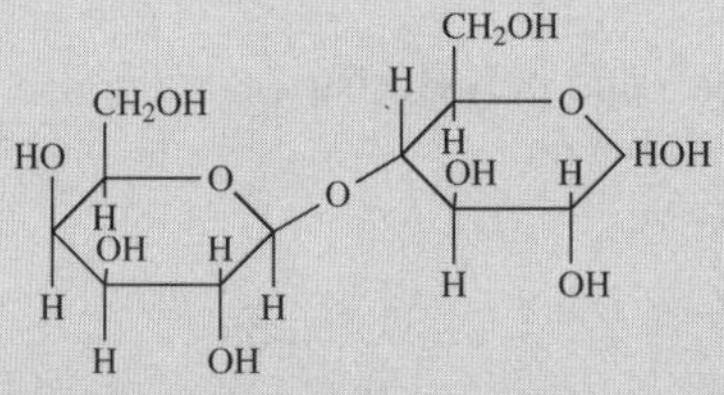

(e) trehalose

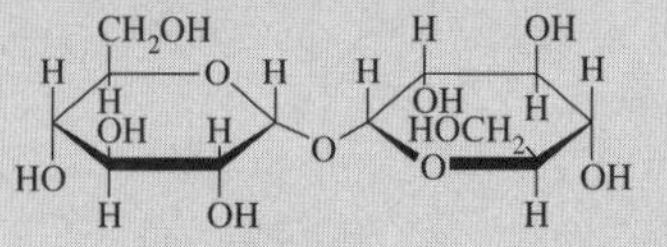

6.4 For the scheme below provide the reagents, **a-d**, and the unknown compounds, **(W)–(Z)**.

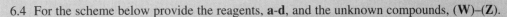

6.5 Illustrating your answer with reference to glyceraldehyde (HOCH$_2$.CHOH.CH=O), explain
   (a) the essential condition for optical activity,
   (b) the meaning of the signs (+) and (−),
   (c) the meaning of the terms D- and L-, and
   (d) the meaning of the terms (R)- and (S).
   How could glyceraldehyde be converted into an aldotetrose?

6.6 Give the Fischer projections of L-sorbose and D-ribose.

6.7 Into which hexoses will D-galactose be transformed in weakly alkaline solution?

6.8 A non-reducing sugar A, $C_{18}H_{32}O_{16}$, on acidic hydrolysis gives D-glucose and D-fructose in a ratio of 2:1. Cautious hydrolysis of A gives D-glucose and $C_{12}H_{22}O_{11}$, a reducing sugar. Give three structures for A which would satisfy this data.

7 Amino acids, peptides and proteins

Topics

7.1 Amino acids
7.2 Peptides and proteins
7.3 Summary

Peptides and proteins are among the most important of biological molecules. For example, peptides are a major constituent of the soft structural tissue of animals; enzymes (proteins) act as catalysts for cell reactions; nucleoproteins control metabolic activity in the cell; soluble proteins in the blood are concerned with oxygen transport (haemoglobin) and immune response; and there are numerous polypeptide hormones.

The importance of peptides and proteins is rivalled only by the nucleic acids (see p. 185), which control heredity, and also direct the synthesis of proteins. Proteins are high molecular weight polymers (polyamides) of the α-amino acids.

7.1 Amino acids

Numerous types of amino acids are known but we will concentrate on the most biologically significant α-amino acids.

7.1.1 Structure

Amino acids are usually drawn with both an amino and a carboxyl group (**1**) but are actually better represented as the dipolar (zwitterionic) structure (**2**) in neutral solution.

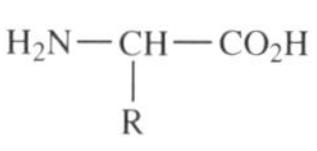

160 Amino acids, peptides and proteins

$$H_2N-CH(R)-CO_2H \qquad \overset{\oplus}{H_3N}-CH(R)-CO_2^{\ominus}$$
(1) 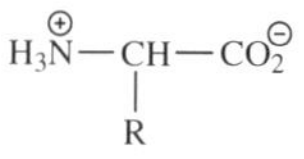 (2)

The presence of both basic and acidic groups means that in an aqueous solution of an amino acid, the species present is pH-dependent. At low pH (<4) the carboxylic acid group will be undissociated and the amino group protonated. At high pH (>10) the amino group will be in the form of an aminocarboxylate anion. At intermediate pH the zwitterion will be the principal species in solution (Table 7.1).

Table 7.1

pH	Predominant species	Net charge
1	RCH($\overset{+}{N}H_3$)CO$_2$H	+1
7	RCH($\overset{+}{N}H_3$)CO$_2^-$	0
14	RCH(NH$_2$)CO$_2^-$	−1

The pH of a solution of an amino acid at which the average charge per molecule is zero is known as the '**isoelectric point**'. (A monoamino monocarboxylic acid has two pK_a values corresponding to dissociation of the $-\overset{+}{N}H_3$ and CO$_2$H groups in the cation, RCH($\overset{+}{N}H_3$)CO$_2$H. For these simple acids the isoelectric point is given by pH$_{\text{isoelectric}}$ = (pK_1 + pK_2)/2.)

Twenty-three amino acids have been found in proteins, all of which are α-amino acids. Being all α-amino acids these species have some properties in common, but others, such as hydrophobicity, acidity/basicity, etc., are dependent upon the other functional groups that are present.

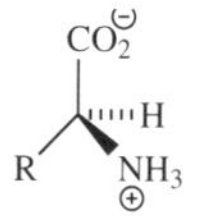

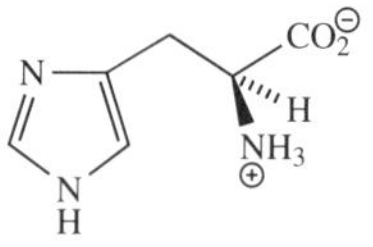

Histidine (H) 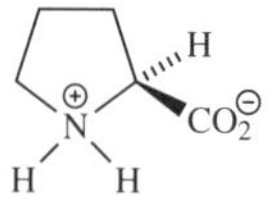   Proline (Pro; P)

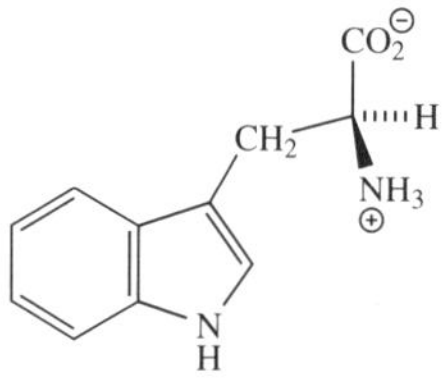

Tryptophan (W) 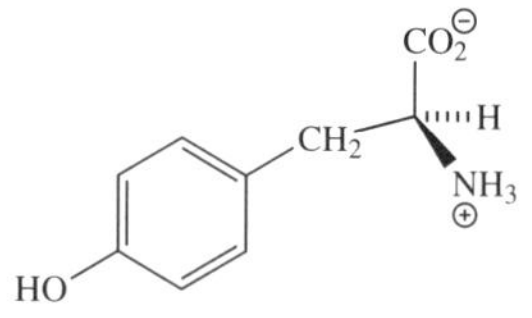   Tyrosine (Y)

Ten of these amino acids are essential, in that they cannot be synthesised by animals and must, therefore, be obtained from the diet (marked with E in Table 7.2).

Table 7.2 Some common amino acids

Amino acid	—R	Symbol	
Alanine	—CH$_3$	Ala	(A)
Arginine[E]	—(CH$_2$)$_3$NHC(=NH)NH$_2$	Arg	(R)
Asparagine	—CH$_2$CONH$_2$	Asn	(N)
Aspartic acid	—CH$_2$CO$_2$H	Asp	(D)
Cysteine	—CH$_2$SH	Cys	(C)
Glutamine	—(CH$_2$)$_2$CONH$_2$	Gln	(Q)
Glutamic acid	—(CH$_2$)$_2$CO$_2$H	Glu	(E)
Glycine	—H	Gly	(G)
Histidine[E]	—CH$_2$(4-imidazolyl)	His	(H)
Isoleucine[E]	—CH(CH$_3$)CH$_2$CH$_3$	Ile	(I)
Leucine[E]	—CH$_2$CH(CH$_3$)$_2$	Leu	(L)
Lysine[E]	—(CH$_2$)$_4$NH$_2$	Lys	(K)
Methionine[E]	—(CH$_2$)$_2$SCH$_3$	Met	(M)
Phenylalanine[E]	—CH$_2$Ph	Phe	(F)
Proline	See previous page	Pro	(P)
Serine	—CH$_2$OH	Ser	(S)
Threonine[E]	—CH(CH$_3$)OH	Thr	(T)
Tryptophan[E]	—CH$_2$(3-indolyl)	Trp	(W)
Tyrosine	—CH$_2$(4-hydroxyphenyl)	Tyr	(Y)
Valine[E]	—CH(CH$_3$)$_2$	Val	(V)

All α-amino acids, except glycine (R = H) contain a chiral centre and can thus exist as a pair of enantiomers. Usually, only one of the pair is found naturally, i.e. in peptides:

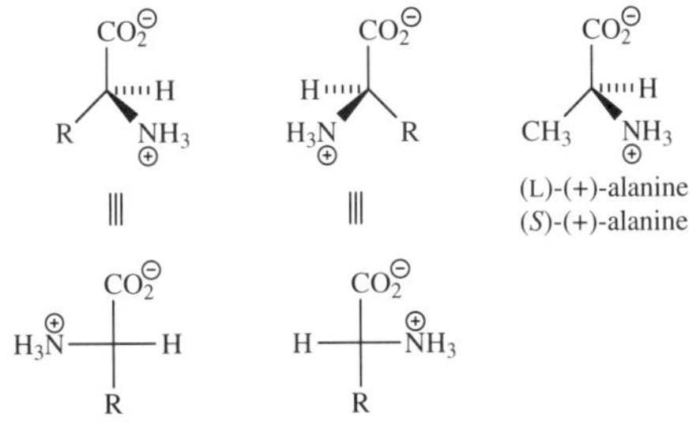

(L)-(+)-alanine
(S)-(+)-alanine

(D/L-nomenclature also based on glyceraldehyde.)

All amino acids from proteins have the (L)-configuration, i.e. $\overset{+}{N}H_3$ on the LHS in Fischer projection, and this mostly means that they also have the (S)-configuration.

7.1.2 Preparation

α-Amino acids can be prepared in the laboratory by the action of ammonia on an α-halocarboxylic acid (nucleophilic substitution), or by hydrolysis of the aminonitrile, which can be derived from the lower aldehyde (the Strecker synthesis). Both of these routes will give racemic products:

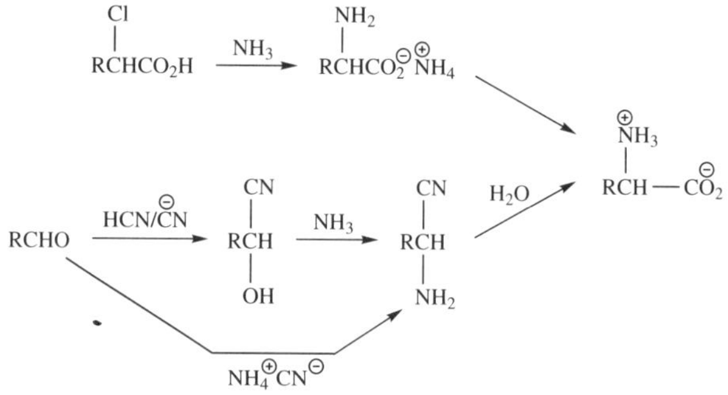

Numerous chiral syntheses of α-amino acids are now known (including an asymmetric version of the Strecker synthesis), with some of the most important being the asymmetric hydrogenation of dehydroamino acids (using a chiral rhodium catalyst) followed by hydrolysis, e.g. the Monsanto process for the preparation of the anti-Parkinson's disease agent, (L)-dopa.

(L)-Dopa (R, R)-DIPAMP

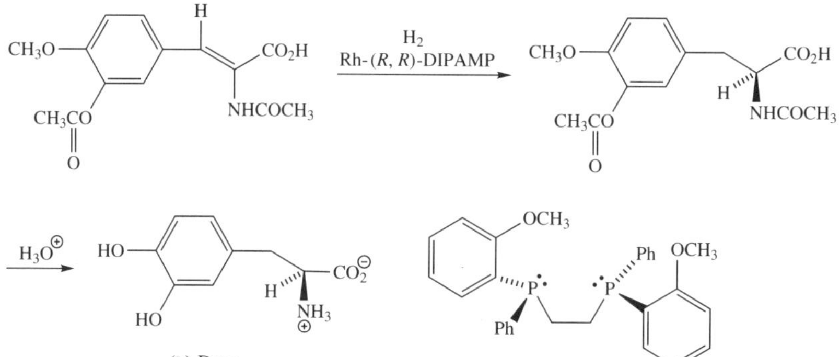

((*R,R*)-DIPAMP is a chiral bisphosphine ligand.)

There are also numerous asymmetric enzymatic syntheses of α-amino acids and one example is the use of (L)-specific aminopeptidase (from *Pseudomonas putida*), which hydrolyses only the (L)-amide from the racemic mixture, to the (L)-acid, leaving the (D)-amide untouched.

(L)-Phenylglycine (D)-Phenylglycine amide

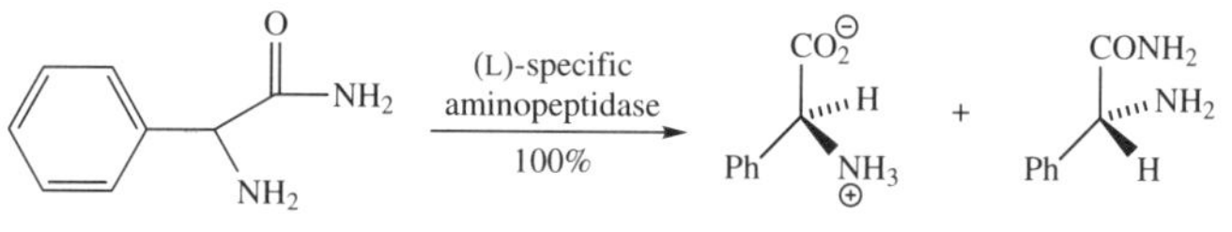

7.1.3 Reactions

Although they exist primarily in the zwitterionic form, the primary amine function of α-amino acids has all the reactions characteristic of such a group. Two important reactions are the reaction with methanol (formaldehyde) to give the corresponding methylene-imines, which, unlike amino acids themselves, can be titrated with alkali using normal indicators, and so are used in the estimation of amino acids; and the enzymic oxidation of α-amino acids to α-keto acids (see p. 129):

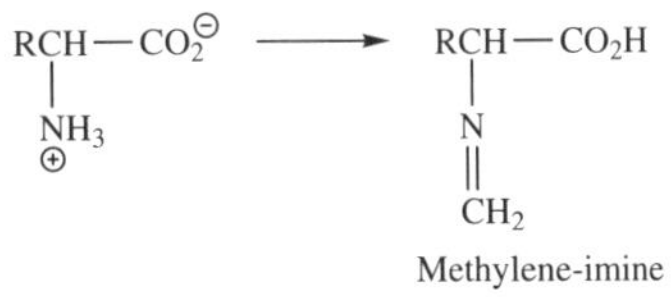

$$\begin{array}{c} RCH-CO_2^{\ominus} \\ | \\ \overset{\oplus}{NH_3} \end{array} \longrightarrow \begin{array}{c} RCH-CO_2H \\ | \\ N \\ \parallel \\ CH_2 \end{array}$$

Methylene-imine

The carboxyl group of α-amino acids also shows some of the normal reactions, forming salts, and chelate complexes which are useful in the isolation of amino acids:

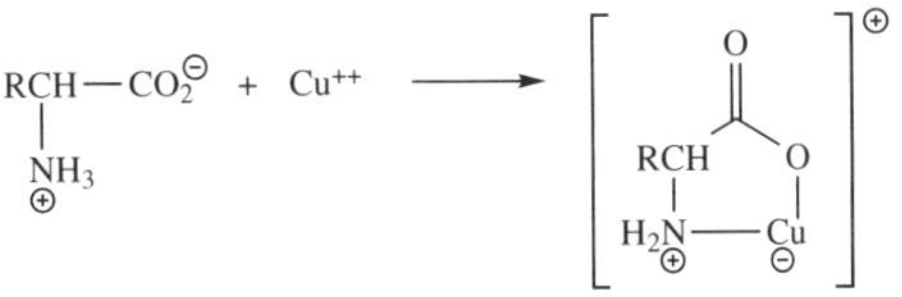

$$\begin{array}{c} RCH-CO_2^{\ominus} \\ | \\ \overset{\oplus}{NH_3} \end{array} + Cu^{++} \longrightarrow \left[\begin{array}{c} O \\ \parallel \\ RCH \quad O \\ | \quad | \\ H_2\overset{\oplus}{N}-\overset{\ominus}{Cu} \end{array} \right]^{\oplus}$$

Esterification is possible in the presence of sufficient mineral acid to ensure that the carboxylate anion is completely protonated.

Primary α-amino acids react with ninhydrin (also known as triketohydrindene hydrate and indanetrione hydrate – the *gem*-diol formed by hydration of a triketone) – to give an intense purple colour. The reaction occurs in two stages, the amino acid being initially oxidised to the lower aldehyde or ketone with liberation of ammonia and carbon dioxide. The ammonia then reacts with the compound produced by reduction of ninhydrin and unchanged ninhydrin to give the purple compound. The quantitative formation of this substance coupled with the intensity of its colour makes this a valuable

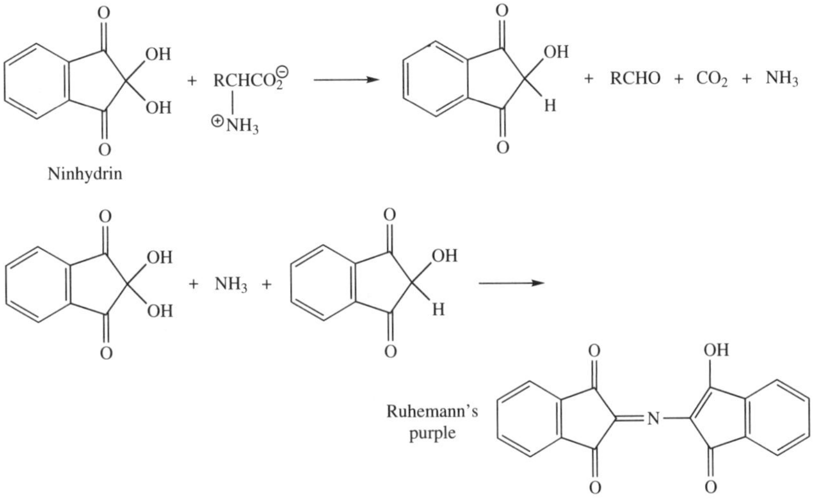

Ninhydrin

Ruhemann's purple

reaction, which has been extensively used for the qualitative detection of amino acids (e.g. as a chromatographic spray) and for quantitative estimation by spectrophotometric means. The intensity of the colour produced makes the method correspondingly sensitive. It should be noted that only primary α-amino acids can give this test with ninhydrin. β-, γ- and δ-Amino acids are not oxidised in the first step, and secondary and tertiary α-amino acids are oxidised but do not liberate ammonia. Secondary α-amino acids such as proline and hydroxyproline give much fainter yellow colours with ninhydrin and cannot be estimated in this way.

7.2 Peptides and proteins

Peptides and proteins are groups of compounds of similar structure, differing only in their molecular size. Both are polyamides, formed by the condensation reaction of the carboxyl group of one α-amino acid with the amino group of another, and have the general structure:

$$\overset{\oplus}{H_3N}-\underset{\underset{R'}{|}}{CH}-CO(NH\underset{\underset{R}{|}}{CH}-CO)_n-NH\underset{\underset{R''}{|}}{CH}-CO_2^{\ominus}$$

The amido group, —NHCO—, is referred to as the **peptide linkage**. Where the polyamide is constructed from relatively few amino acids, the term 'peptide' is used, with a prefix to indicate the number of amino acid residues. The term 'protein' is usually used where the number of amino acids is very large, and proteins are known that have relative molecular masses up to 10^4 or 10^6; however, there is no clearly defined boundary between the terms 'peptide' and 'protein'.

$$\overset{\oplus}{H_3N}-CH_2-\overset{\overset{O}{\|}}{C}-NHCH_2CO_2^{\ominus}$$
glygly
(dipeptide)

$$\overset{\oplus}{H_3N}-\underset{\underset{CH_2CONH_2}{|}}{CH}-CONH\underset{\underset{CH_2SH}{|}}{CH}CONHCH_2CO_2^{\ominus}$$
asncysgly
(tripeptide)

N-terminal amino acid residue $\overset{\oplus}{H_3N}-\underset{\underset{R}{|}}{CH}-\overset{\overset{O}{\|}}{C}(NH\underset{\underset{R}{|}}{CH}C)_n NH\underset{\underset{R}{|}}{CH}CO_2^{\ominus}$ C-terminal amino acid residue

Polypeptide

The entire amide group is flat (N p-orbital and those forming C=O π bond must be parallel to allow for delocalisation) and the C—N bond thus has considerable double bond character and restricted rotation:

Amides are non-basic since the lone-pair is delocalised and therefore not available for donation. The double bond character of the C—N bond means that the amide bond is relatively stable to hydrolysis (cf. C—O bond of esters). Hydrolysis can be achieved by enzymic or 'chemical' means. Hot dilute mineral acid will slowly cleave the amide links to give random degradation, ultimately resulting in the formation of simple amino acids. Controlled acidic hydrolysis will degrade a protein into a mixture of peptides. Enzymic hydrolysis is also possible, and the proteolytic enzymes vary greatly in their specificity. Some, like papain or ficin, are virtually non-specific and degrade proteins to free amino acids, whilst others like trypsin, chymotrypsin and pepsin will hydrolyse only particular links in protein molecules (cf. maltase, emulsin, etc., p. 150). Thus pepsin will cleave the amide link between the carbonyl group of a dicarboxylic L-amino acid and the amino group of an aromatic L-amino acid, provided that the second carboxylic acid group of the former is uncombined. Chymotrypsin is less specific and cleaves the amide link on the carbonyl side of aromatic L-amino acids. Trypsin is specific for amide links involving the carboxyl group of lysine or arginine. In all these

Pepsin cleaves here — Chymotrypsin cleaves here

$$—NH—CH(—(CH_2)_n—CO_2H)—CO—NH—CH(—CH_2—C_6H_4—X)—CO—NH—CH(R)—CO—$$

$n = 1$ (Aspartic acid) or 2 (Glutamic acid)

$X = H$ (Phenylalanine) or OH (Tyrosine)

cases the enzyme not only requires the presence of a particular amino acid, but will hydrolyse the link on only one side of the amino acid, not on both. The specificity of these enzyme-catalysed reactions undoubtedly involves interaction of the side-chain functionality of the polypeptide with receptor sites on the proteolytic enzyme. Other enzymes with differing specificity are known: carboxypeptidases degrade peptides or proteins with a free terminal carboxylic acid group by stepwise removal of single amino acids, and aminopeptidases effect a similar stepwise degradation from the end with a free amino group. Neither of these groups of enzymes is effective in the hydrolysis of peptides or proteins with a cyclic structure.

The investigation of the structure of proteins utilises these and other methods of degradation, and a number of techniques have been devised to identify the terminal amino acids. One, widely employed to identify the amino-terminal amino acid of a chain, is to react the polypeptide with 2,4-dinitrofluorobenzene, which converts the free amino group into its 2,4-dinitrophenyl derivative. Subsequent hydrolysis of the polypeptide gives normal amino acids, with the exception of the terminal *N*-aryl amino acid, which can be separated and identified chromatographically.

The primary structures of two biologically important peptides are given below. Glutathione is a tripeptide, required as a cofactor in some enzymic

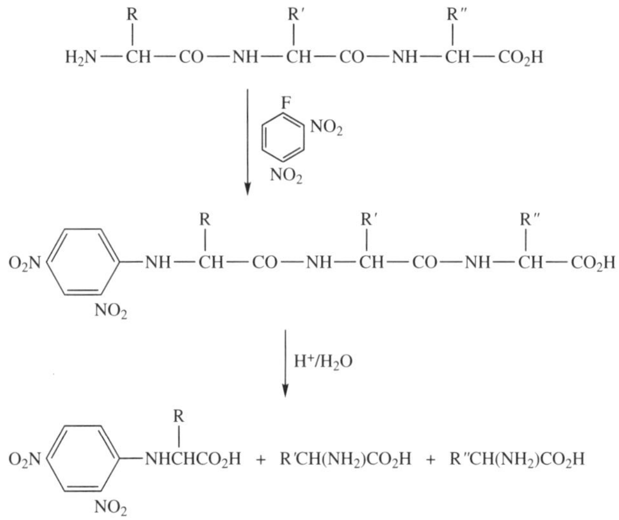

166 Amino acids, peptides and proteins

$$H_2N-CH(R)-CO-NH-CH(R')-CO-NH-CH(R'')-CO_2H$$

$$\downarrow \text{F-C}_6\text{H}_3(\text{NO}_2)_2$$

$$O_2N-C_6H_3(NO_2)-NH-CH(R)-CO-NH-CH(R')-CO-NH-CH(R'')-CO_2H$$

$$\downarrow H^+/H_2O$$

$$O_2N-C_6H_3(NO_2)-NHCH(R)CO_2H \;+\; R'CH(NH_2)CO_2H \;+\; R''CH(NH_2)CO_2H$$

oxidations. It is unusual in having the glutamic acid residue linked through the γ-carboxylic acid function.

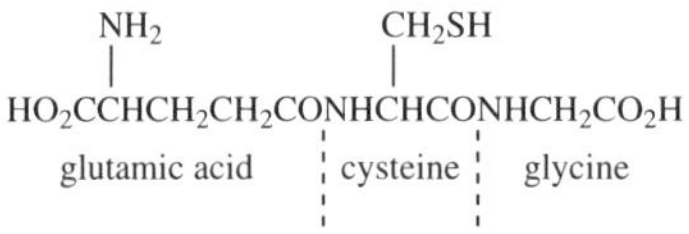

Glutathione: $HO_2CCH(NH_2)CH_2CH_2CONHCH(CH_2SH)CONHCH_2CO_2H$
(glutamic acid — cysteine — glycine)

Oxytocin structure showing: cystine, tyrosine, isoleucine, asparagine, glutamine, proline, leucine, glycinamide.

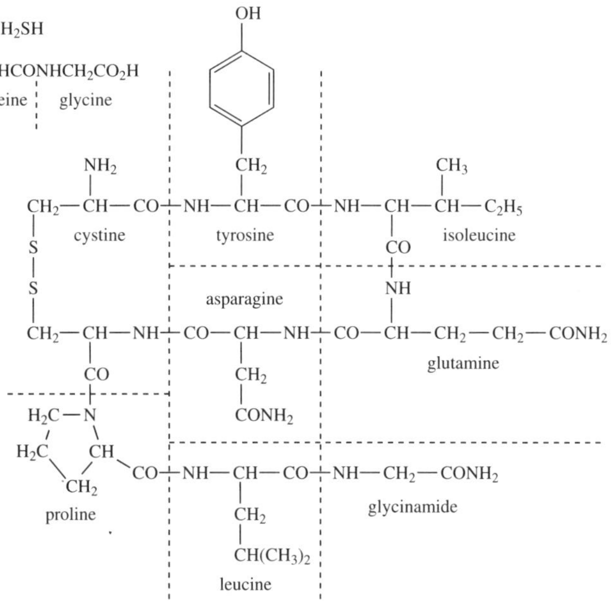

Oxytocin

Oxytocin is a more complex octapeptide hormone secreted by the posterior lobe of the pituitary gland, which stimulates contraction of the uterus and initiates lactation at the end of the pregnancy. A very similar octapeptide hormone vasopressin, also secreted by the posterior pituitary, is the anti-diuretic hormone, and differs from oxytocin in the replacement of L-isoleucine by L-phenylalanine, and L-leucine either by L-lysine (in hog vasopressin) or

L-arginine (in beef vasopressin). The primary structure of beef insulin, which contains 51 amino acid residues, is shown below. The amino-terminal end of the peptide chain is shown by the symbol H · (e.g. H · Gly · below means H$_2$NCH$_2$CO—) and the carboxylic acid end by · OH (e.g. · Ala · OH means —NHCH(CH$_3$)CO$_2$H).

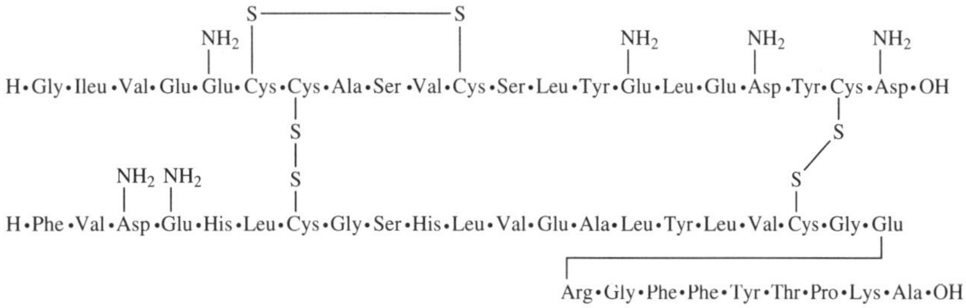

```
                    S─────────────S
             NH₂    │             │              NH₂    NH₂    NH₂
              │     │             │               │      │      │
H·Gly·Ileu·Val·Glu·Glu·Cys·Cys·Ala·Ser·Val·Cys·Ser·Leu·Tyr·Glu·Leu·Glu·Asp·Tyr·Cys·Asp·OH
                              │                                              │
                              S                                              S
                              │                                             /
       NH₂ NH₂                S                                             S
        │   │                 │                                             │
H·Phe·Val·Asp·Glu·His·Leu·Cys·Gly·Ser·His·Leu·Val·Glu·Ala·Leu·Tyr·Leu·Val·Cys·Gly·Glu
                                                                                  │
                                           Arg·Gly·Phe·Phe·Tyr·Thr·Pro·Lys·Ala·OH
```

Beef insulin (for key to abbreviations see Table 7.2)

7.2.1 The structure and physical properties of proteins

Much chemical and biochemical interest is focused upon the relationship between the structure and function of proteins. Peptides and proteins may have both basic (NH$_2$, CO$_2^-$) and acidic ($\overset{+}{N}$H$_3$, CO$_2$H) groups in the molecule either as the terminal groups of the polyamide chain or due to inclusion of polyfunctional amino acids such as lysine or glutamic acid. The net charge on the molecule will vary with the pH of its environment, just as with the simple amino acids (p. 159). Thus, under conditions of low pH the protein will be positively charged, but at high pH it will be negatively charged. At some intermediate pH the net or average charge on the molecule will be zero and this pH is referred to as the **isoelectric point** (cf. p. 160).

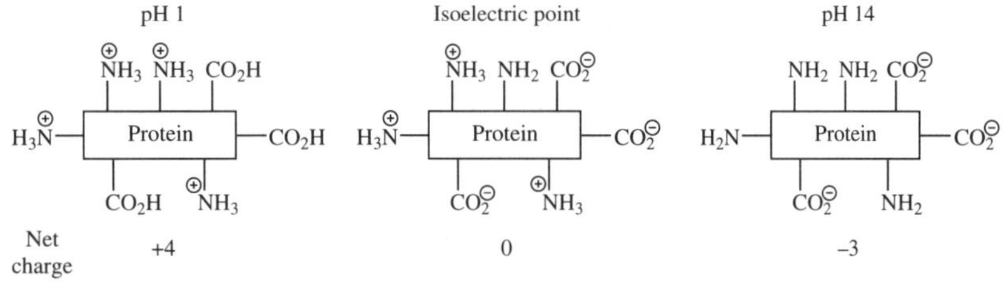

	pH 1	Isoelectric point	pH 14
	$\overset{\oplus}{N}$H$_3$ $\overset{\oplus}{N}$H$_3$ CO$_2$H	$\overset{+}{N}$H$_3$ NH$_2$ CO$_2^{\ominus}$	NH$_2$ NH$_2$ CO$_2^{\ominus}$
	H$_3$$\overset{\oplus}{N}$—[Protein]—CO$_2$H	H$_3$$\overset{\oplus}{N}$—[Protein]—CO$_2^{\ominus}$	H$_2$N—[Protein]—CO$_2^{\ominus}$
	CO$_2$H $\overset{\oplus}{N}$H$_3$	CO$_2^{\ominus}$ $\overset{\oplus}{N}$H$_3$	CO$_2^{\ominus}$ NH$_2$
Net charge	+4	0	−3

It is not very surprising that the properties of protein solutions show marked changes on passing through the isoelectric point, since the solvation of molecules or interaction of adjacent protein molecules is clearly likely to be affected by the charge distribution and total charge on the molecule. The viscosity of gelatin solution passes through a minimum at pH 4.7 (the isoelectric point) and the solubilities of insulin and casein are lowest at their isoelectric points (5.3 and 4.7 respectively).

The variation of the total charge on a polypeptide or protein with change in the pH of the medium can be used to separate these molecules by the technique of **electrophoresis**. If a mixture of polypeptides in an aqueous buffer solution of known pH is subjected to a powerful electric field, molecules with overall positive and negative charges will migrate in opposite directions, while those with net zero charge at the chosen pH will remain stationary. By conducting the operation on a buffer-impregnated paper or in a film of conducting jelly, a complex mixture of proteins can be separated into its components. Changes in the pH at which this process is conducted will change the mobility or direction of migration of the constituents of the mixture.

The activity of many enzymes is known to be pH-dependent, activity being at a maximum at a particular pH. It is known that enzymes adsorb their substrates at a special region of the molecule known as the 'active site'. Changes in pH will change the distribution of charges on the molecule, which in its turn will affect hydration, either by varying the number of hydrogen-bonding groups or by changing the extent to which water molecules cluster around the protein by dipolar interaction with charged sites. The actual receptor groups of the active site, which bind the substrate, may also be protonated or deprotonated. All these effects may reduce the ease with which the enzyme adsorbs its particular substrate and so decrease the catalytic activity.

Quite apart from the effect of environment upon the biological activity of proteins, it is known that a protein's structure is intimately related to its function. It is customary to divide the structural features of proteins into a number of categories. The **primary structure** of a protein is the sequence of amino acid residues, as determined by chemical analysis, and this chain may be coiled or arranged in a particular fashion as a result of hydrogen bonding between amide groups. Each H-bond imparts 20–40 kJ mol^{-1} of stabilisation energy so the most stable protein structure will be that with the maximum possible H-bonds. Those features of the protein structure that result from such interactions between peptide (amide) links are classed as **secondary structure**. Further folding of the secondary structure may result from the interaction of functional groups in the side chains of the amino acids (—SH, NH$_2$, CO$_2$H, —OH, etc.) and this constitutes the **tertiary structure**. Finally, proteins may have a **quaternary structure** resulting from the interaction of several protein molecules to give groups or clusters which may possess a high degree of symmetry, sometimes directly observable by the electron microscope.

Numerous polypeptides and proteins have been studied by X-ray crystallography, and some features of their structure have been recognised. Two types of regular secondary structure are found to occur frequently, although numerous more disorderly arrangements are also known (Figure 7.1). In the 'α' form, the polyamide chain is coiled into a helix, in which adjacent turns of the coil (approximately 3.6 amino acid residues per turn) are joined by hydrogen bonds between neighbouring amide groups. The helix is dextral, which ensures that the bulky side chains of the L-amino acids project away from the spiral polyamide chain. In the 'β' structure the polyamide chains are arranged adjacent in a parallel or an anti-parallel fashion to form a sheet of polypeptide chains cross-linked by hydrogen bonds. In this array the side chains of the amino acids lie alternately above and below the plane of the sheet. Sections of

Peptides and proteins 169

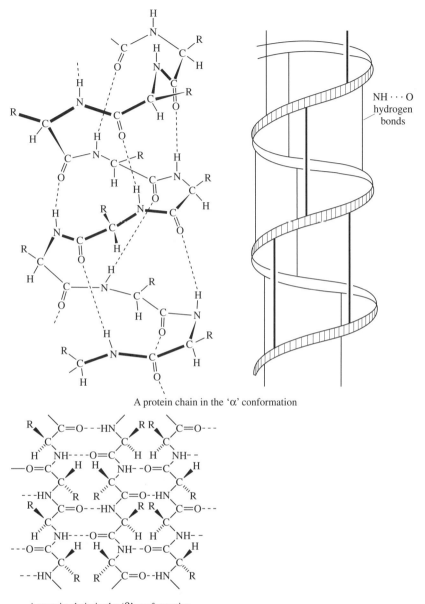

Figure 7.1   A protein chain in the 'β' conformation

β structure can be formed within a single protein molecule by pleated folding of the polypeptide chain. Proteins that have extensive β structure, such as silk fibroin, are not readily extensible, since the polypeptide chains are already fully extended, but those with the α structure predominating (e.g. hair, wool) are elastic, since mechanical stress can be relieved by conversion of the helical polypeptide chain into the extended conformation of the β structure.

The tertiary structure of proteins, involving interaction of the side chains of amino acids, does not lead to regularity such as described above. Besides hydrogen bonding, the formation of disulphide links is an important factor in the stabilisation of tertiary structure. Insulin has three such disulphide bridges, two of which link the two separate polypeptide chains in the molecule. The tertiary structure often holds the protein molecule in a conformation in which the hydrophilic groups (such as OH, NH_2 and CO_2H) are exposed on the surface and the hydrophobic groups (alkyl and aryl side chains) are enclosed within the centre of the molecule.

The quaternary structure of proteins varies widely. Some electron micrographs clearly reveal the aggregation of protein molecules, whose fine structure cannot be resolved. One common form of quaternary structure found in fibrous proteins (wool, hair) consists of six protein chains, each individually in the form of an α helix, being coiled around a central helical protein molecule to give a rope-like structure.

The biological activity of proteins is often intimately connected with the higher orders of structure, and living organisms are able to synthesise proteins in the desired conformations, which frequently are metastable (i.e. not the most stable structure possible). Under the influence of heat, extremes of pH, or many chemical reagents, proteins will often lose their biologically desirable conformation, being converted into random and disoriented structures with consequent loss of biological activity. This process, known as denaturation, is most familiar in the change of texture of egg white on heating, and is also responsible for the change in texture of meat during cooking. In the latter case, cooking causes a marked increase in the digestibility of the meat, since denaturation exposes linkages in the protein, which in the raw state are not so readily accessible to the proteolytic enzymes of the digestive tract. In these denaturations, unfolding of the protein chains exposes the hydrophobic groups normally enclosed within the central part of the molecular shape, and interaction between the exposed hydrophobic sections of neighbouring molecules contributes to the coagulation of denatured protein.

Although of no biological significance, the processes of hair-waving exemplify interference with secondary and tertiary structure in different ways. Water-waving utilises the ability of water to penetrate the protein tissue, which is softened by the breaking of hydrogen bonds between amide groups in the protein and formation of new hydrogen bonds to water molecules. On drying, new hydrogen bonds are reformed within the protein, which preserves the shape resulting from external constraint. In permanent waving a similar result is achieved by the reduction of disulphide bridges to thiol groups, followed by reoxidation (by setting lotion) to form a new set of disulphide links.

7.3 Summary

1. Amino acids are best represented in their zwitterionic (dipolar) form and the pH of a solution of an amino acid at which the average charge is zero is known as the 'isoelectric point'.

$$\underset{R}{\overset{CO_2^{\ominus}}{\underset{NH_3^{\oplus}}{\bigg|}}}\!\!\!\!\text{''}H \quad \text{L-Amino acid}$$

2. All α-amino acids, except glycine, are chiral since they have four different groups attached to the α-carbon atom. All α-amino acids from proteins are L, and this usually means that they are also the (S)-enantiomers.

3. α-Amino acids can be prepared by a number of methods, including hydrolysis of an aminonitrile and the action of ammonia on an α-halocarboxylic acid. Numerous asymmetric syntheses are known, including the use of enzymes, or chiral catalysts.

4. Peptides and proteins are polyamides, with the amide group being known as the peptide link:

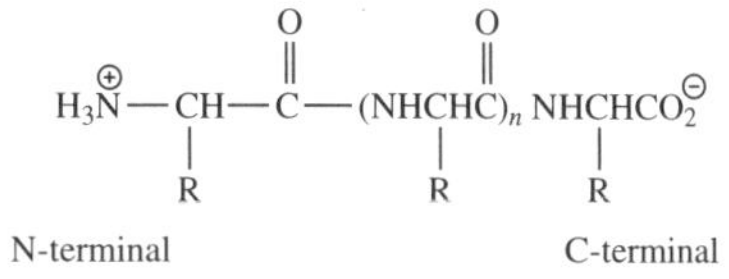

$$H_3\overset{\oplus}{N}-CH-\underset{R}{\overset{O}{\overset{\parallel}{C}}}-(NHCHC)_n\underset{R}{NHCHCO_2^{\ominus}}$$

N-terminal   C-terminal

Chemical hydrolysis of peptides/proteins cleaves all peptide links but enzymic hydrolysis is often selective.

5. The primary structure of a protein is the sequence of amino acid residues and plays a major role in the chemical properties of the protein. The secondary structure of proteins arises from hydrogen bonding between peptide links, whereas the tertiary structure arises from the interaction of functional groups on the side chains. Finally, the quaternary structure results from the interactions of several protein molecules.

Problems

7.1 Which L-amino acid does not have the (S) configuration? (Hint: consider the priority of the groups attached to the α-carbon atom.)

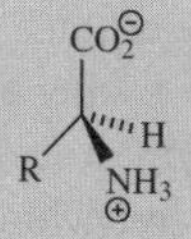

7.2 How would glutathione and oxytocin be affected by pepsin, chymotrypsin and aminopeptidase?

7.3 A compound $C_{12}H_{17}N_3O_3$ on acidic hydrolysis gives L-tyrosine, L-alanine and ammonia. Give all possible structures for this compound. How might you attempt to distinguish between these structures?

8 Aromatic compounds, nucleic acids and nucleotide coenzymes

Topics

8.1 Aromatic compounds
8.2 Nucleic acids
8.3 Nucleotide coenzymes
8.4 Summary

8.1 Aromatic compounds

Benzene, C_6H_6, is the simplest of a large number of highly unsaturated, cyclic or polycyclic hydrocarbons. Benzene and its derivatives were originally classified as aromatic owing to their distinctive odours but nowadays the term aromatic has a more rigorous scientific meaning. In benzene the carbon atoms form a regular hexagon which is planar and all C—C bond lengths are the same (0.139 nm – almost intermediate between a C=C double bond, 0.133 nm, and a C—C single bond, 0.147 nm). Each carbon is bonded to one hydrogen and all C—H bond lengths are the same, 0.109 nm. All bond angles in benzene are 120°. Thus, all the C atoms in benzene are equivalent. They are sp^2 hybridised (hence the bond angles of 120°) and the hexagon is composed of a σ bonded framework formed by the overlap of sp^2 orbitals on C with sp^2 orbitals on the adjacent C, or with the s orbital on H. This leaves one electron in an unhybridised p orbital on C, see (**1**) in Figure 8.1. These p orbitals interact to form extended π orbitals, which stretch around the ring (cf. conjugation, p. 11), and the six electrons occupy three of these orbitals, of which the simplest consists of two circular electron clouds, one above and one below the ring of carbon atoms (**2**). The other orbitals, of which there are a total of six

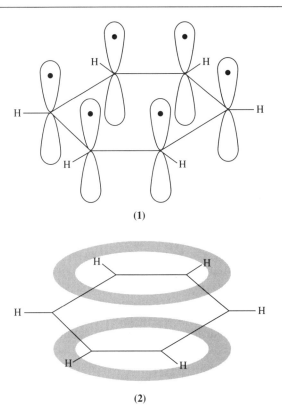

Figure 8.1

but only the lowest three (bonding) are filled, have a somewhat different symmetry but in all of them electrons can move freely around the six carbon atoms. The characteristic chemical properties of aromatic compounds, which will be discussed later, are attributable to the presence of these cyclic orbitals.

The chemical behaviour of benzene (and other aromatics) is distinct from that of alkenes, as we shall see later, but one reaction that aromatics and alkenes have in common is reduction and this gives us an opportunity to measure the so-called resonance energy of aromatics.

8.1.1 Resonance energy of benzene

Upon reduction of both but-2-ene and cyclohexene, to butane and cyclohexane respectively, the enthalpy change is -119 kJ mol^{-1}:

$$\text{CH}_3\text{CH}=\text{CHCH}_3 \text{ (g)} + \text{H}_2 \text{ (g)} \longrightarrow \text{CH}_3\text{CH}_2\text{CH}_2\text{CH}_3$$
But-2-ene → Butane

$$\Delta H^\ominus (298 \text{ K}) = -119 \text{ kJ mol}^{-1}$$

Cyclohexene (l) + H_2 (g) ⟶ Cyclohexane (l)

$$\Delta H^\ominus (298 \text{ K}) = -119 \text{ kJ mol}^{-1}$$

Hence, if benzene consisted of three non-interacting C=C double bonds, we would expect that hydrogenation would give an enthalpy change of $-357\,\text{kJ}\,\text{mol}^{-1}$. The actual value, however, is $-208\,\text{kJ}\,\text{mol}^{-1}$:

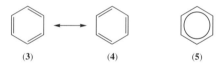

$\Delta H^{\ominus}(298\,\text{K}) = -208\,\text{kJ}\,\text{mol}^{-1}$

The difference between these values ($149\,\text{kJ}\,\text{mol}^{-1}$) is the **resonance energy** – the energy released when three isolated double bonds interact to form the cyclic delocalised molecular orbitals. Any reaction that results in the destruction of the cyclic molecular orbitals will require the restoration of the lost $149\,\text{kJ}\,\text{mol}^{-1}$ and so will be energetically unfavourable. This loss of resonance stabilisation explains why aromatic compounds undergo substitution reactions, which preserve the aromatic character, unlike alkenes which normally undergo addition reactions (see p. 44).

Several methods of representing the benzene ring in structural formulae are in use, of which only the Kekulé structures (**3**) and (**4**) will be employed in this book, although it should be noted that these do not imply alkene-like double bonds. Another common representation (**5**) has serious disadvantages in terms of electron counting in mechanisms, and will not be used here.

8.1.2 Other aromatic systems

Benzene is only one of a large number of compounds that exhibit certain characteristic chemical and physical properties. The common feature of these compounds is that they contain a planar, cyclic, conjugated π system of $(4n + 2)$ electrons.

Of the carbocyclic (i.e. a ring of carbon atoms) systems the cyclopropenium cation is the simplest aromatic system, having a 2π electron ($n = 0$) system. The next simplest aromatic systems have 6π electrons ($n = 1$) and include benzene, the cyclopentadienyl anion and the tropylium cation. The formation of an aromatic system is energetically very favourable and, as a result, cyclopentadiene forms a very stable aromatic anion upon reaction with alkali metals. In fact, cyclopentadiene ($pK_a \sim 16$) is almost as acidic as water ($pK_a = 15.75$). Monocyclic systems with larger numbers of π electrons are also known, e.g. [18] annulene.

Planar, cyclic systems with $(4n)$ π electrons are said to be **anti-aromatic** and are very reactive and unstable compared with their non-planar

6 (*n* = 1) [Cyclopentadienide anion resonance structures] ⟷ ⟷ ⟷ etc.

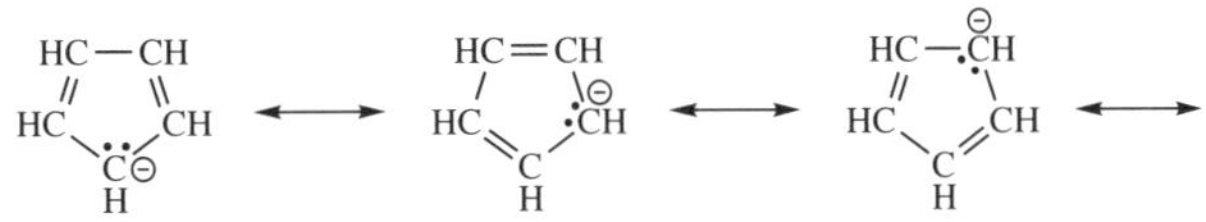

Cyclopentadienide anion

6 (*n* = 1) [Tropylium cation resonance structures] ⟷ ⟷ ⟷ etc.

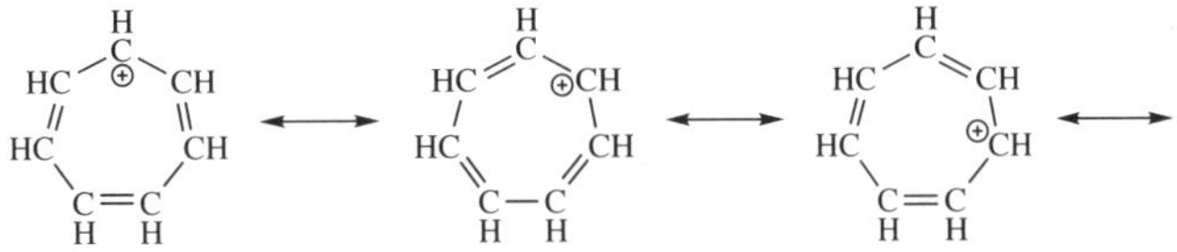

Tropylium cation

18 (*n* = 4) [[18] Annulene structure]

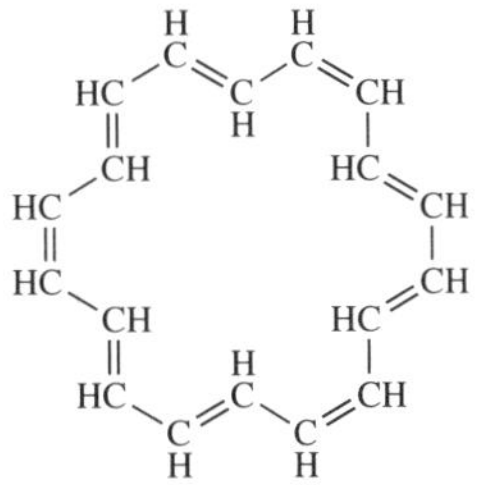

[18] Annulene

conformations. Cyclobutadiene (4π electrons) is very reactive and cyclooctatetraene (8π) is non-planar, and shows alkene-like reactions:

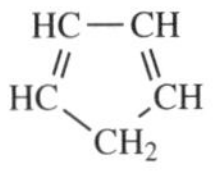

Cyclopentadiene 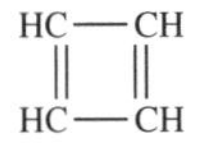   Cyclobutadiene

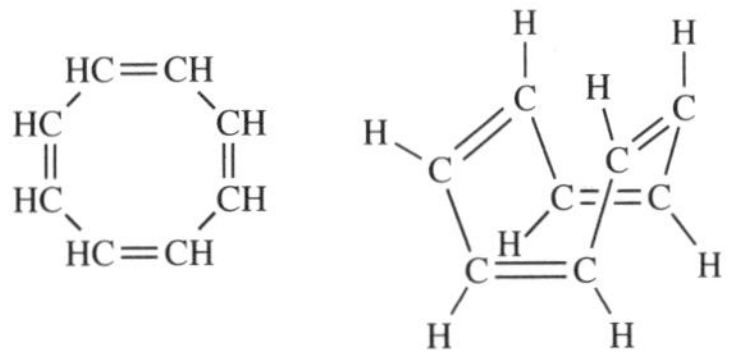

Cyclooctatetraene

A large number of aromatic compounds are derived from systems of 'fused' benzene rings. The simpler members of this series, naphthalene, anthracene and phenanthrene, have typical aromatic properties, and several Kekulé-type structures can be drawn for these molecules. Some of the higher polycyclic aromatic hydrocarbons, such as benzpyrene, are powerful carcinogens, the presence of which in oils (e.g. shale oil) has been responsible for certain occupational cancers.

Fused polycyclic aromatic hydrocarbons

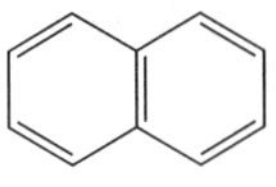

Naphthalene 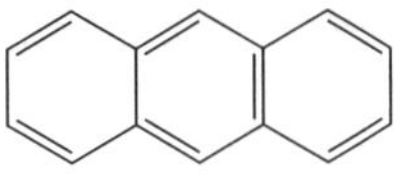   Anthracene 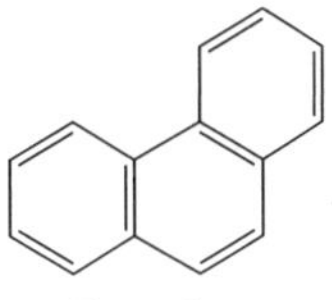   Phenanthrene

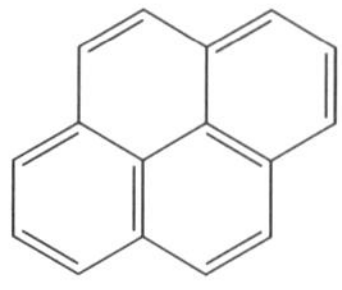

Pyrene 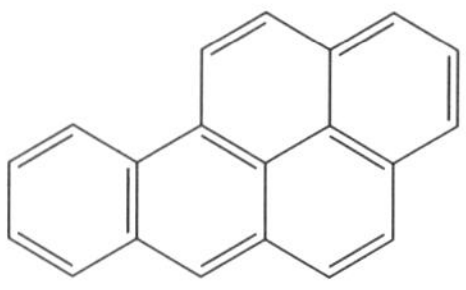   4, 5-Benzpyrene

Benzenoid heterocyclic aromatic systems

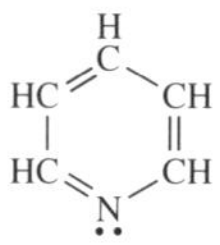

Pyridine 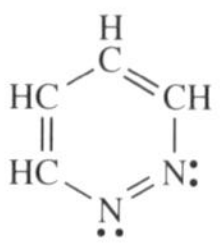   Pyridazine 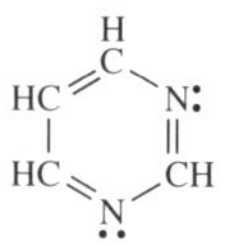   Pyrimidine 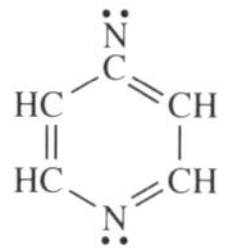   Pyrazine

Non-benzenoid heterocyclic aromatic systems

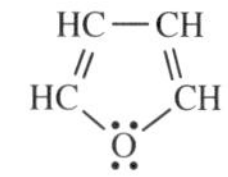

Furan 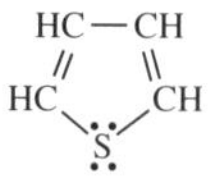   Thiophene 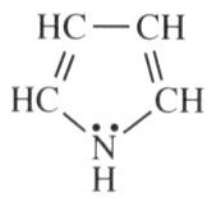   Pyrrole 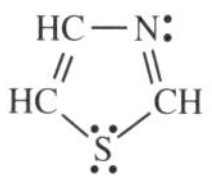   Thiazole 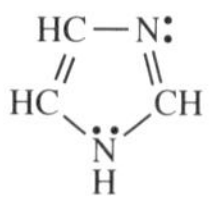   Imidazole

(In the diagrams above, lone-pairs contributing to the aromatic sextet of electrons are shown inside the ring, lone-pairs not involved in the aromatic sextet are shown outside.)

Heterocyclic systems

Aromaticity is not restricted to carbocyclic compounds, and replacement of some of the carbon atoms in the compounds mentioned above by other atoms gives new aromatic systems, provided that the π electron system is unchanged. Replacement of CH groups in benzene by the isoelectronic (i.e. containing the same number of electrons) nitrogen atom results in a series of heterocyclic aromatic compounds: pyridine, pyridazine, pyrimidine and pyrazine, and even further replacement is possible. In all these compounds the cyclic 6π electron system – the 'aromatic sextet' – utilises one of the electrons from each carbon and nitrogen atom, leaving a lone-pair in an sp^2 orbital, at right angles to the p orbitals, on each nitrogen, where in benzene there would be a C—H bond. As a result these heterocyclic compounds are feebly basic, the basicity of a lone-pair in an sp^2 orbital being markedly less than that of a lone-pair in an sp^3 orbital (cf. the acidity of C—H in alkanes and alkynes, p. 107). The cyclopentadienide anion can also be regarded as the carbocyclic parent of a series of heterocyclic aromatic compounds. Furan and thiophene have an aromatic sextet in which one electron comes from each of the four carbon atoms (i.e. the two double bonds) and two are contributed from a lone-pair on oxygen or sulphur. In pyrrole, the lone-pair on the nitrogen atom is used in this way, so that pyrrole is non-basic, having no lone-pair on nitrogen available for combination with a hydrogen ion (cf. pyridine where the lone-pair is not involved in the aromatic sextet). Further replacement of the carbon atoms of these rings leads to more complex heteroaromatic species such as imidazole (occurring in the amino acid histidine, p. 160) and thiazole. Note that in thiazole the lone-pair of the nitrogen atom is in an sp² orbital corresponding to a C—H bond in thiophene and is therefore not involved in the aromatic sextet. Thiazole is therefore basic. In imidazole the two nitrogen atoms have quite different utilisation of their electrons, one is basic, as in pyridine, and the other is non-basic, as in pyrrole, since its lone-pair is part of the aromatic sextet.

As with benzenoid (benzene-like) aromatic systems, so with non-benzenoid systems: fused polycyclic aromatic compounds are possible. Quinoline and isoquinoline are derived by fusion of benzene and pyridine rings. Indole, the parent from which the amino-acid tryptophan (p. 161) is derived, consists of benzene and pyrrole rings fused together, and pteridine, purine and alloxazine are more complex examples of heteroaromatic systems obtained in this way. Alloxazine may be considered as a heterocyclic analogue of anthracene.

Fused heterocyclic aromatic systems

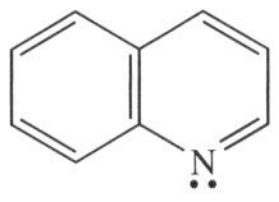

Quinoline 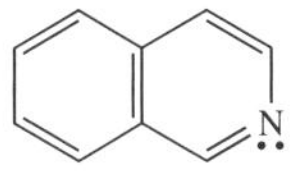   Isoquinoline 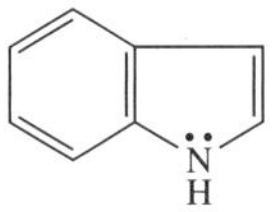   Indole

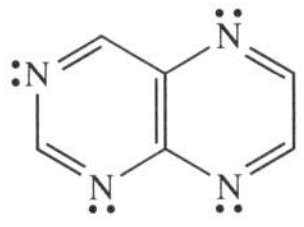

Pteridine 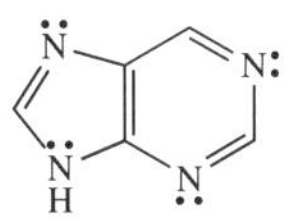   Purine 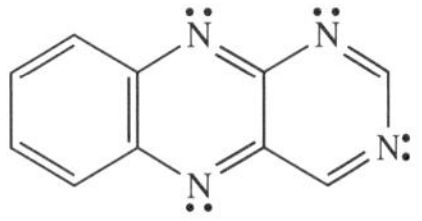   Alloxazine

Although the parent heterocyclic compounds do not occur naturally, their derivatives are widespread in nature and of considerable importance. Nicotinamide, the amide of nicotinic acid, and pyridoxal (Vitamin B_6) are both derived from pyridine, and both are vitamins of the B group. Nicotinamide is an important part of the coenzymes NAD and NADP (p. 191), and pyridoxal

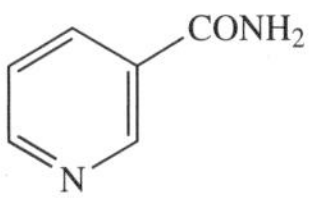

Nicotinamide 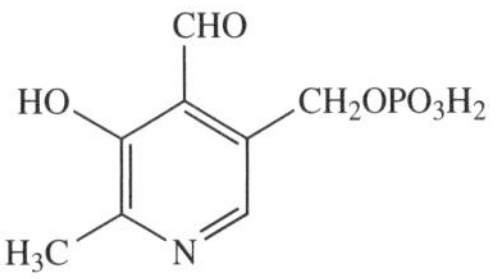   Pyridoxal phosphate

phosphate is a cofactor required for decarboxylation and transamination of amino acids. Pyrimidine bases are significant as part of the nucleic acid structure (p. 185) and are also found in a number of coenzymes. Vitamin B_1

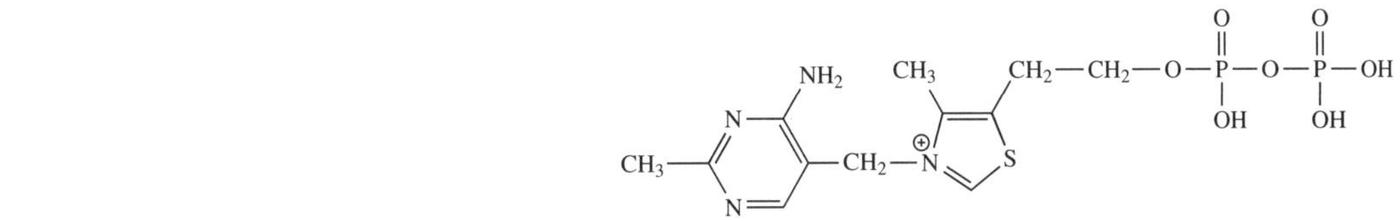

Thiamine pyrophosphate

(thiamine) is a derivative of pyrimidine and thiazole which is required, as its pyrophosphate, as a cofactor in the enzymic decarboxylation of α-keto-acids. Simple pyrazine and pyridazine derivatives are of no biological significance, but the pteridine system is found in folic acid (Vitamin B_{10}):

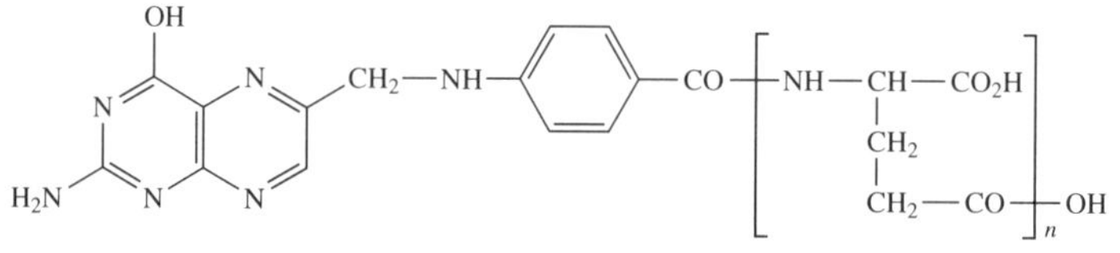

Folic acid (n = 2–6 depending upon the source)

The tetrahydro-derivative of folic acid (see below) is required as a cofactor in biological syntheses in which single carbon units undergo oxidation or reduction and transfer onto substrate molecules as —CH_3, —CH_2OH or —CHO groups. Riboflavin (Vitamin B_2) is derived from alloxazine and occurs as the prosthetic group (the non-protein portion of an enzyme, which is frequently involved in the enzymically catalysed reaction) in a number of dehydrogenases (enzymes which oxidise by removal of hydrogen):

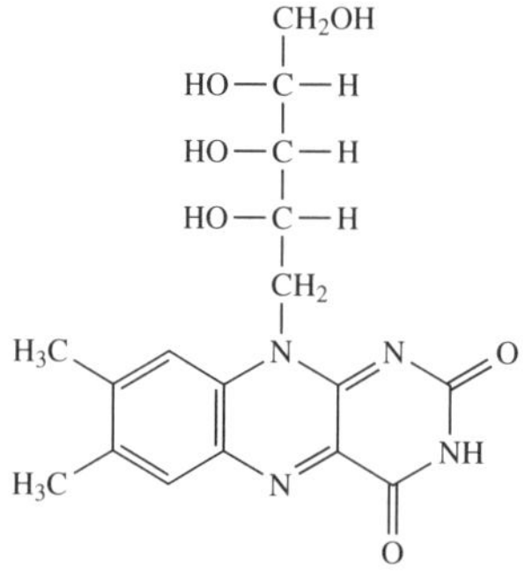

Riboflavin

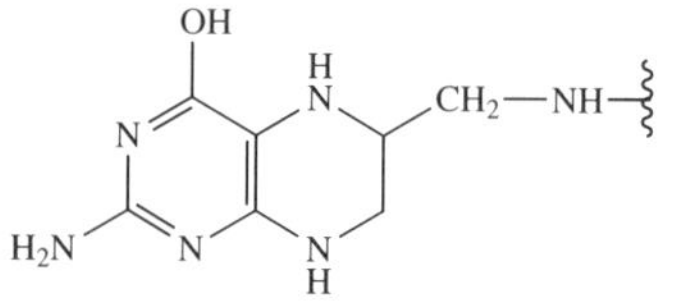

Tetrahydrofolic acid (part structure)

Purine derivatives are important in the nucleic acids, and purine is the skeleton of uric acid, the principal end product of nitrogen metabolism in terrestrial invertebrates, birds and reptiles. Caffeine, the principal stimulant and diuretic of coffee, is a simple purine derivative, and theophylline and theobromine which occur in tea and cocoa respectively have very similar structures.

Many of the drugs used in medicine are derivatives of these and other heteroaromatic systems.

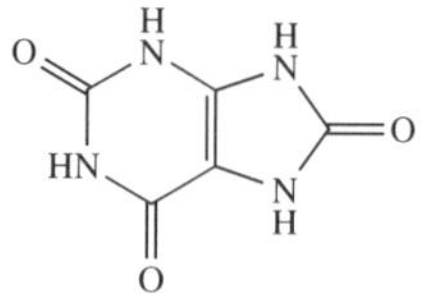

Uric acid

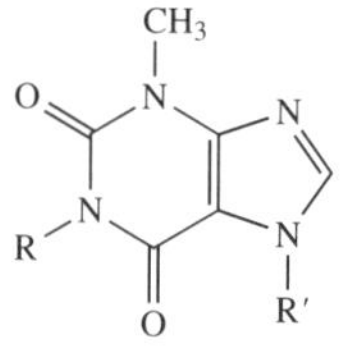

Caffeine R = R' = CH_3
Theophylline R = CH_3 R' = H
Theobromine R = H R' = CH_3

8.1.3 Electrophilic substitution

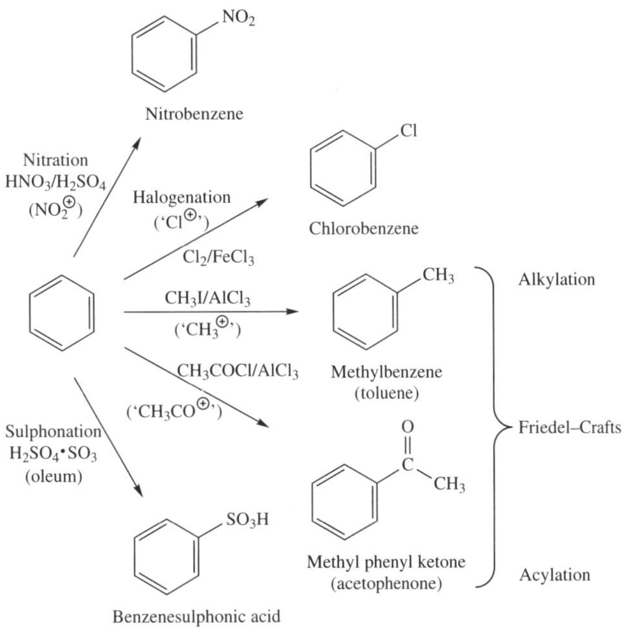

The reactions that distinguish aromatic compounds from alkenes are electrophilic substitutions, of which the most important are:

- **nitration**, by a mixture of nitric and sulphuric acids;
- **halogenation**, by chlorine or bromine in the presence of a Lewis acid catalyst, e.g. $FeCl_3$, $AlCl_3$;
- **Friedel–Crafts alkylation or acylation**, in which alkyl or acyl halides, in the presence of $AlCl_3$, substitute the aromatic ring with alkyl or acyl groups respectively;
- **sulphonation** by concentrated sulphuric acid or oleum.

Although these reactions may seem diverse in type, they are mechanistically very similar. In all cases, the reaction is initiated by attack by the electron-rich (nucleophilic) benzene ring on an electrophile, E⁺, to form an intermediate carbocation (**6**) known as a Wheland intermediate or σ complex. This carbocation intermediate is resonance-stabilised since the positive charge can be delocalised around the ring. Loss of a proton from the carbocation forms the substituted aromatic system (**7**).

The difference between the reactions of alkenes and aromatic hydrocarbons lies in the second step. Whereas the carbocation formed in the electrophilic attack on an alkene promptly reacts with an anion (Y⁻), if the corresponding reaction occurred with the carbocation (**6**), the product (**8**) would no longer be aromatic. To convert an aromatic compound to a non-aromatic one requires the return of the 149 kJ mol⁻¹ of resonance energy lost in forming the aromatic π orbitals (p. 173). Addition reactions of aromatic compounds akin to those of alkenes are, therefore, energetically very unfavourable, resulting in the easier alternative of substitution.

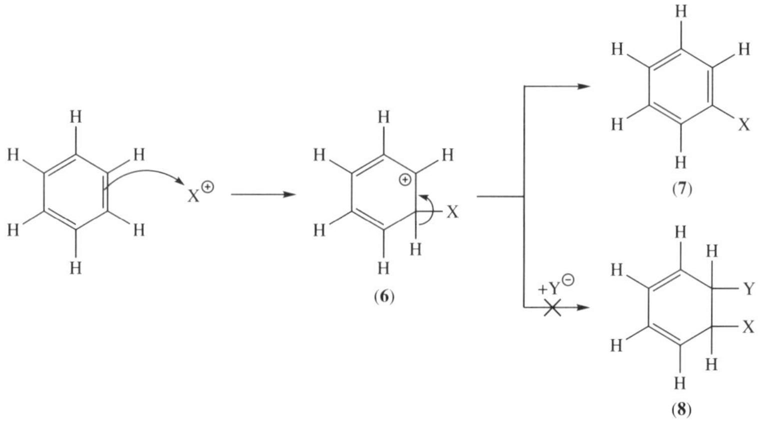

A great deal is known about the reactive species involved in the electrophilic substitutions listed above. **Nitration** has been studied extensively and is known to occur via attack by the nitronium ion NO_2^+. This is produced by the nitrating mixture as follows:

$$H-\ddot{O}-NO_2 + H_2SO_4 \rightleftharpoons H-\overset{+}{\underset{H}{O}}-NO_2 + HSO_4^-$$

$$H-\overset{+}{\underset{H}{O}}-NO_2 \rightleftharpoons H_2O + \overset{+}{N}O_2$$

The role of sulphuric acid is not to remove the water formed during nitration, as nitration is rarely reversible. Any strong acid (e.g. $HClO_4$) that will protonate nitric acid, and thereby generate nitronium ions, will serve in its place. Since salts of the nitronium ion are known as crystalline solids (e.g. $NO_2^+ClO_4^-$; $NO_2^+BF_4^-$; $NO_2^+NO_3^-$ (solid N_2O_5)), and can be used in nitrations, there is no reason to doubt its existence.

The electrophilic species involved in **halogenation** and the **Friedel–Crafts** reactions are less well established than the nitronium ion. In these reactions the electrophile is probably formed via metal halide complexes, which can either generate a cationic species or act directly as the electrophile:

$$Cl_2 + FeCl_3 \rightleftharpoons Cl^+[FeCl_4^-]$$

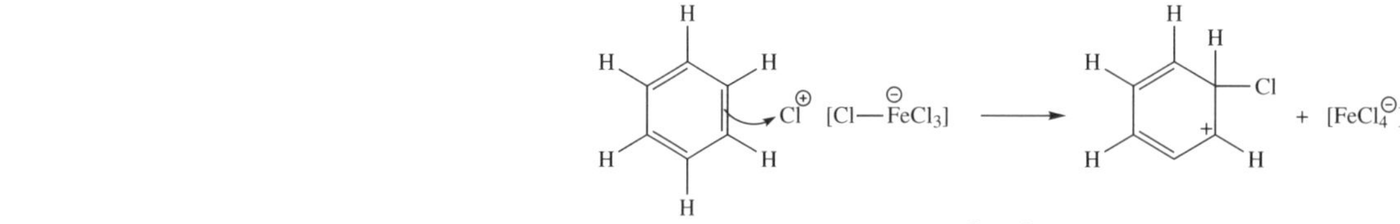

$$R-Cl + AlCl_3 \rightleftharpoons R-\overset{+}{Cl}-\overset{-}{AlCl_3} \rightleftharpoons R^+ + [AlCl_4]^-$$

$$R-CO-Cl + AlCl_3 \rightleftharpoons R-CO-\overset{+}{Cl}-\overset{-}{AlCl_3} \rightleftharpoons R-\overset{+}{CO} + [AlCl_4]^-$$

Sulphonation occurs via attack by sulphur trioxide, which is itself a powerful electrophile:

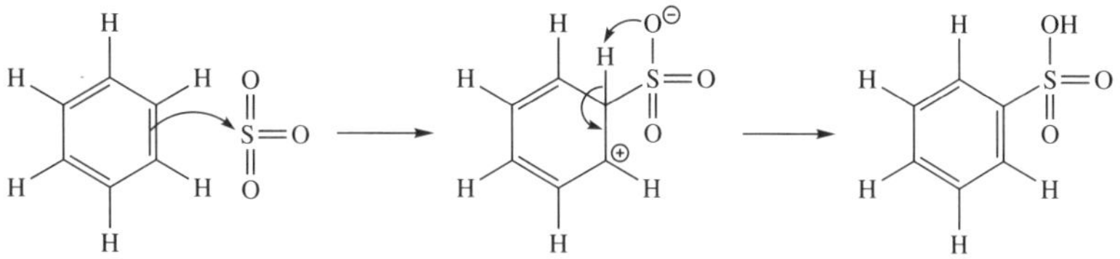

The orientation of substitution

It is convenient to consider, at this stage, the effect of substituents on the position of electrophilic attack. The well-established chemical equivalence of the six carbon atoms of benzene means that there is only one isomer of any monosubstituted benzene, but further substitution can lead to three isomeric disubstituted products:

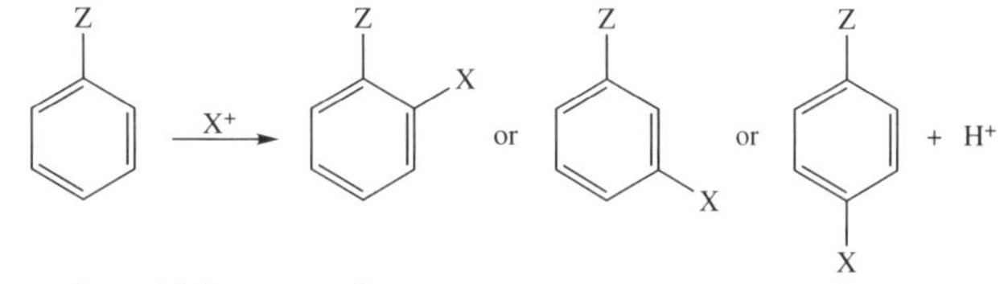

(X^+ = any electrophile) ortho- meta- para-isomer

It is found that the position of attack of the electrophile E^+ is affected by the nature of Z in a way that can be explained by consideration of the structure of the intermediate carbocation. When an electrophile becomes attached to the aromatic system, the carbocation formed is a resonance hybrid of three canonical structures (p. 13) and is stabilised by the distribution of positive charge over several atoms. This is sometimes indicated by diagrams such as (**9**). When considering the introduction of a second substituent, there will be

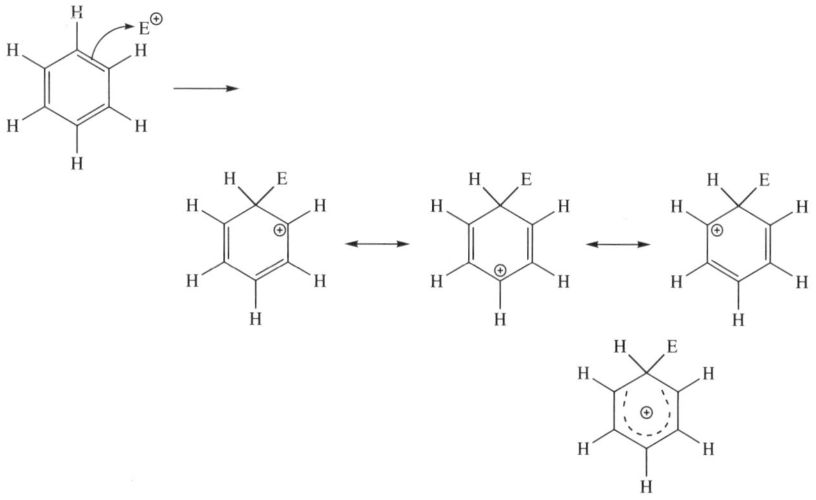

(**9**)

three canonical structures for each of the intermediate carbocations involved in *ortho*, *meta*- and *para*-substitution, and these are shown in diagrams (**10**) to (**18**). It should be noted that in the case of *ortho*- and *para*-substitution, canonical structures can be drawn for the intermediate in which the positive charge resides on the carbon atom bearing the substituent group Z (**12**, **18**), while no such structure is possible for the intermediate involved in *meta*-substitution. It is the interaction of the group Z with the carbocation which determines which intermediate is energetically favoured, and therefore which isomer is ultimately obtained. It should be realised that quite small changes in the energy of one of the canonical structures can have a pronounced effect upon the ease of formation of the carbocation. Substituents can be classified into three groups.

Activating; ortho/para directing. In this case, the substituent Z is either electron-donating or has a lone-pair on the atom adjacent to the benzene ring. These groups activate the benzene ring towards electrophilic aromatic substitution since they increase the electron density on the ring, making it more nucleophilic and, hence, facilitate the first step in the mechanism, the attack on the electrophile. Benzene rings substituted with such substituents are more prone to undergo electrophilic attack. Electron-donating substituents (CH_3, C_2H_5, etc.) can stabilise structures (**12**) and (**18**) by inductive displacement of electrons towards the adjacent carbon atom (**19**). The result of this interaction is that the carbocation intermediates in *ortho*- and *para*-substitution are more easily formed than the intermediate of the *meta*-substitution path, in which no such interaction is possible. This type of group therefore leads to predominant *ortho*- and *para*-substitution by electrophiles, with weak activation of the benzene ring.

If the substituent Z has a lone-pair on the atom adjacent to the benzene ring then a mesomeric interaction (resonance) can occur, resulting in a fourth resonance form for the carbocations involved in *ortho*- and *para*-substitution

(**21**, **22**), with no corresponding increase in the resonance forms for the *meta*-substituted intermediate. Remembering that an increase in the number of canonical structures means a marked decrease in the energy of formation (i.e. an increase in the ease of formation) of the carbocation, this results in an overwhelming predominance of *ortho*- and *para*-substitution by electrophiles. These substituents (e.g. —NH_2, —OH (strong), —NHCOR, OCOR (moderate)) also result in an increase in the ring electron density and so activate the ring.

Deactivating; meta directing. In this case, the substituent Z (e.g. NO_2, $\overset{+}{N}R_3$, CX_3 (strong), —CO_2R, CHO, COR, CN (moderate)) is more electronegative than carbon and so electron displacement from carbon to Z decreases the electron density of the ring, deactivating it to electrophilic aromatic substitution. This electron withdrawal tends to increase the positive charge on the carbon atom in structures (**20**) thus destabilising these canonical structures and making the carbocations more difficult to form. *Meta*-substitution, which avoids this unfavourable interaction in its intermediate, is, therefore, favoured and predominates.

Deactivating; ortho/para directing. The only substituents that fall into this class are the halogens (—F, —Cl, —Br, —I). Electron withdrawal by these substituents deactivates the ring towards electrophilic aromatic substitution. However, the halogens have lone-pairs and can thus stabilise the carbocation intermediates for *ortho*- and *para*-substitution via a mesomeric interaction (**21**, **22**), leading to a fourth resonance form.

Table 8.1 lists most of the common substituent groups and their effects of electrophilic aromatic substitutions.

8.1.4 Reactions of substituents

We will concentrate here on only some of the important reactions of substituents on an aromatic ring.

184 Aromatic compounds, nucleic acids and nucleotide coenzymes

Table 8.1

Substituent group Z	Activating (A)/ deactivating (D)	Position of electrophilic attack on C_6H_5Z
—Alkyl, e.g. —CH_3	A (weak)	o, p
—CH_2Cl	A (weak)	o, p
—Ph	A (weak)	o, p
—CCl_3	D (strong)	m
—CH=CH_2	A (weak)	o, p
—CHO	D (moderate)	m
—CO.R	D (moderate)	m
—COOH	D (moderate)	m
—COOR	D (moderate)	m
—$CONH_2$	D (moderate)	m
—NH_2	A (strong)	o, p
—NHR	A (strong)	o, p
—NR_2	A (strong)	o, p
—$\overset{+}{N}R_3$	D (strong)	m
—NO_2	D (strong)	m
—OH	A (strong)	o, p
—OR	A (moderate)	o, p
—SH	A (strong)	o, p
—SR	A (moderate)	o, p
—SO_3H	D (moderate)	m
—C≡N	D (moderate)	m
—I, Br, Cl, F	D (weak)	o, p
—NHCOR	A (moderate)	o, p

1. **Oxidation**. Alkylbenzenes are oxidised by $KMnO_4$ (under alkaline conditions) or $Na_2Cr_2O_7$ (under acidic conditions) to give the corresponding benzoic acid:

$$\text{4-nitrotoluene} \xrightarrow{K_2Cr_2O_7 / H_2SO_4} \text{4-nitrobenzoic acid}$$

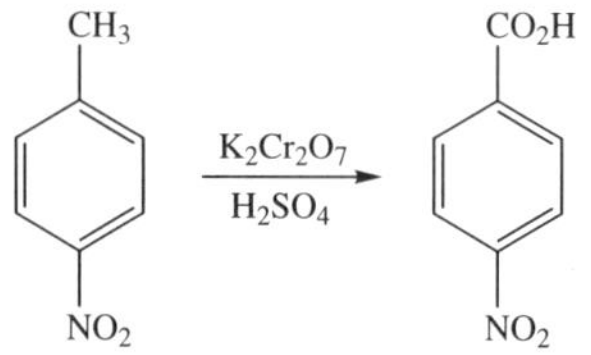

(ethyl and larger side chains also oxidise to the corresponding benzoic acid).

2. **Halogenation**.

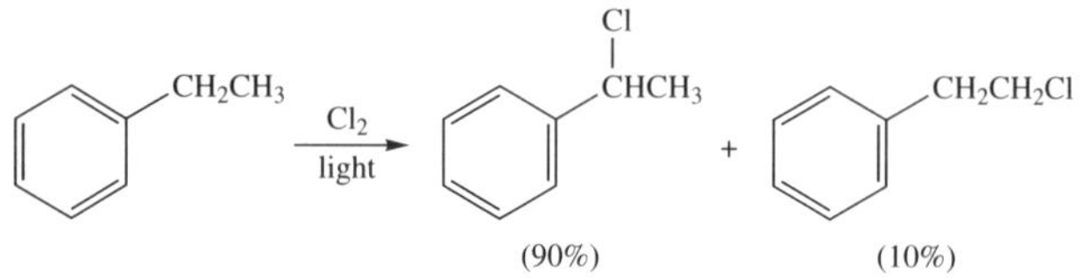

$$\text{PhCH}_2\text{CH}_3 \xrightarrow{Cl_2, \text{light}} \text{PhCHClCH}_3 \,(90\%) + \text{PhCH}_2\text{CH}_2\text{Cl} \,(10\%)$$

In the presence of light a radical halogenation of alkyl side chains occurs (for the mechanism see p. 40).

3. **Formation and reactions of diazonium salts** (see p. 132).

4. **Reduction of aromatic nitro groups**. Aromatic nitro compounds are useful synthetic intermediates as they can be reduced to amines (see p. 125).

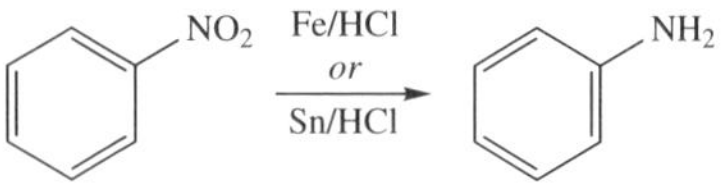

$$C_6H_5NO_2 \xrightarrow[\text{Sn/HCl}]{\text{Fe/HCl or}} C_6H_5NH_2$$

8.2 Nucleic acids

Nucleic acids are acidic macromolecules ('macro' = large) found originally in the nucleus of cells but occurring also in the cytoplasm. Nucleic acids occur in combination with protein, and it is known that viruses, which in some cases can be obtained as crystalline substances, are largely nucleoprotein. The biochemical role of nucleic acids and nucleoproteins is outside the scope of this text, but their importance is shown by the fact that they are responsible for the transmission of hereditary characteristics and control of protein synthesis in the cell.

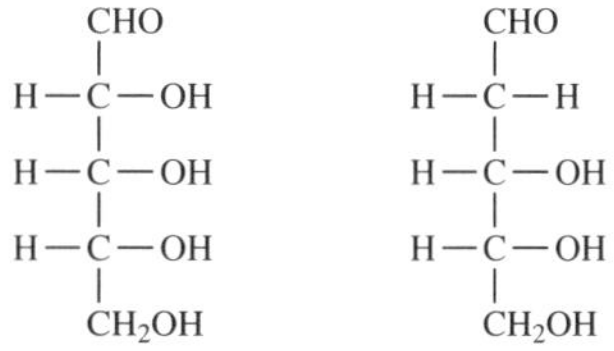

```
    CHO                CHO
    |                  |
H—C—OH             H—C—H
    |                  |
H—C—OH             H—C—OH
    |                  |
H—C—OH             H—C—OH
    |                  |
   CH₂OH              CH₂OH
  D-Ribose         2-Deoxy-D-ribose
```

Separation of the nucleic acids from other cell constituents gives the purified acids as fibrous precipitates. Hydrolysis of purified nucleic acid gives three types of product, these being a group of four basic compounds, a sugar and phosphoric acid. Two types of nucleic acid are known, distinguishable primarily by the sugar obtained on hydrolysis. Ribonucleic acid (RNA) gives

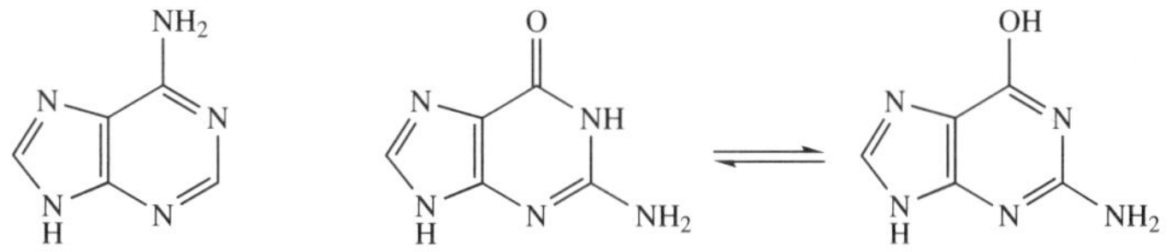

Adenine (A)   Guanine (G)
               (both tautomers)

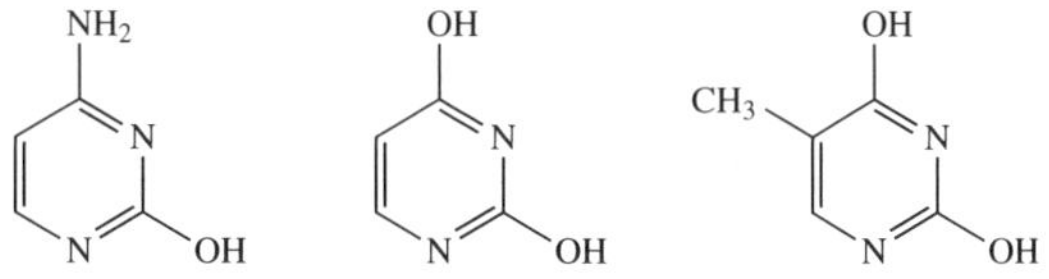

Cytosine (C)   Uracil (U)   Thymine (T)

(Tautomers also possible for C, U and T)

D-ribose, whilst deoxyribonucleic acid (DNA) gives 2-deoxy-D-ribose. The group of bases obtained also varies, both RNA and DNA giving adenine, guanine (derived from purine, p. 177) and cytosine (a pyrimidine), but RNA contains uracil as its fourth base while DNA contains thymine. Enzymic hydrolysis breaks the nucleic acids down into fragments known as nucleotides (containing one molecule each of base, sugar and phosphoric acid), which can be further hydrolysed to nucleosides (composed of one molecule of base combined with one molecule of sugar). This and much other information leads to the conclusion that the nucleic acid consists of a chain of alternate sugar and phosphate residues, with the bases attached to the sugar units:

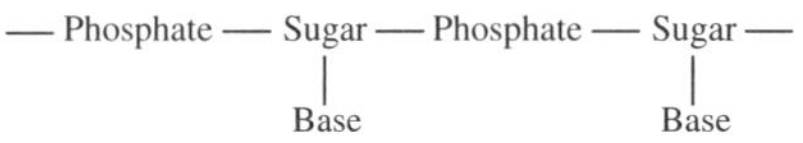

$$\text{— Phosphate — Sugar — Phosphate — Sugar —}$$
$$\qquad\qquad\qquad |\qquad\qquad\qquad\qquad |$$
$$\qquad\qquad\quad \text{Base}\qquad\qquad\qquad\text{Base}$$

The structure of the repeating unit in this chain can be studied in the nucleotides, which have been shown to contain the sugar in a furanose ring (p. 142). Thus the structure of the ribonucleotides derived from adenine and cytosine are:

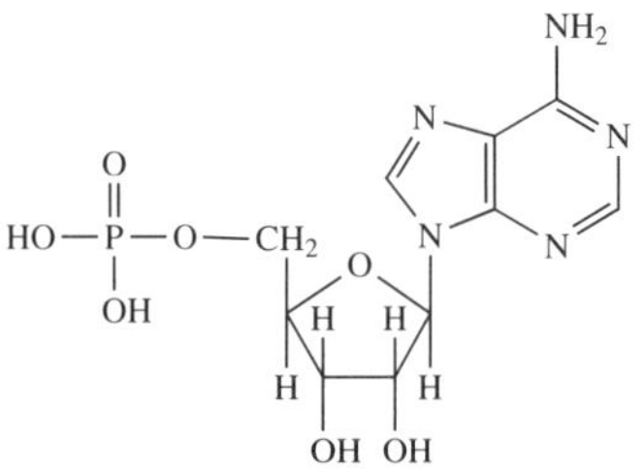

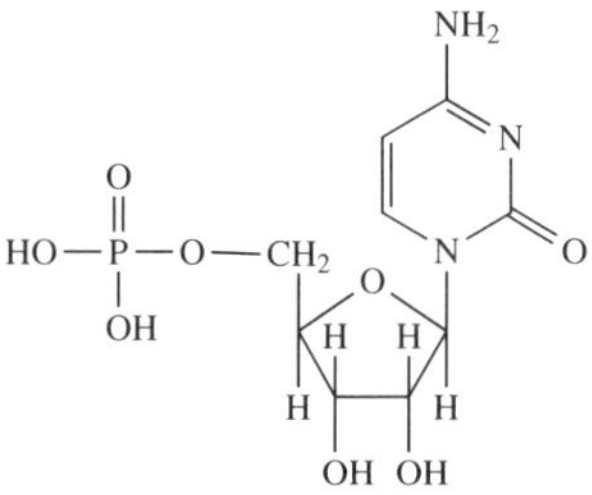

Thus a sequence of the four possible units of the chain in RNA would have the structure given on the opposite page and the whole molecule may contain up to 10^5 such units. A similar sequence is found in DNA, with 2-deoxy-D-ribose as the sugar.

The nucleotide unit containing the characteristic base thymine has the structure:

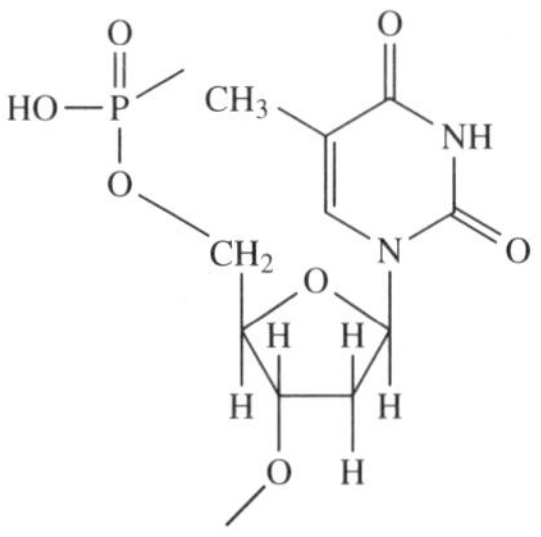

The acidity of these molecules is attributable to the hydroxyl group attached to the phosphorus atom.

Nucleic acids, like proteins, have secondary structure which is important in the biological function, and it is known that DNA exists as a double helix

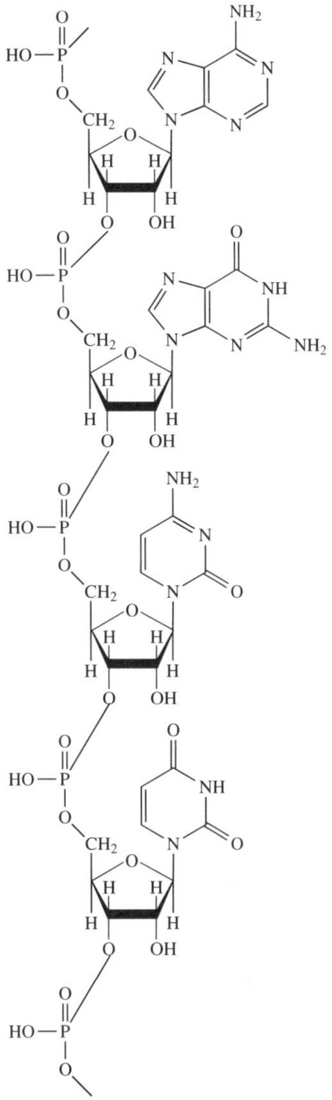

Nucleic acids 187

resulting from the association of two separate nucleic acid molecules. These two molecules associate in a 'head to tail' fashion as defined by the direction of the phosphate link between positions 3 and 5 of adjacent sugar residues:

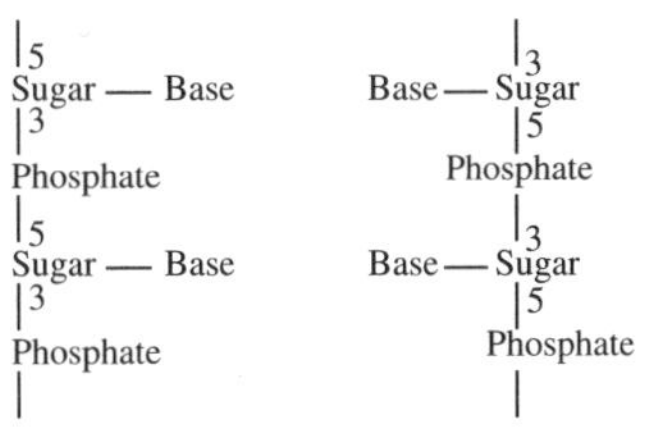

```
     |5                        |3
     Sugar — Base         Base — Sugar
     |3                        |5
     Phosphate              Phosphate
     |5                        |3
     Sugar — Base         Base — Sugar
     |3                        |5
     Phosphate              Phosphate
     |                         |
```

Head-to-tail relationship of two DNA molecules in the double helix

188 Aromatic compounds, nucleic acids and nucleotide coenzymes

In any sample of DNA there is a close correspondence between the abundance of adenine and thymine on the one hand and guanine and cytosine on the other. It is known that the sugar phosphate helix is so arranged that the pyrimidine and purine bases are on the inside of the helix, hydrogen bonding between bases opposed on the two intertwined chains then helps to hold the double helix together, and the pairing of the small pyrimidine bases with the larger purine bases results in the equivalent abundance (Figure 8.2). However, although hydrogen bonding will undoubtedly contribute to stabilisation of the double helix, the strength of binding of the two strands is too great to be accounted for solely in this way.

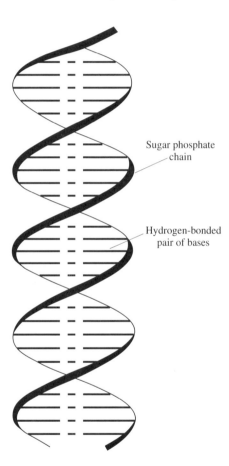

Figure 8.2

RNA molecules do not associate in pairs to form a similar double helix. The bulk of the extra hydroxyl group in the sugar units appears to restrict the conformational flexibility of the polynucleotide chain, and this hydroxyl group must be principally responsible for the ability of enzymes to differentiate between DNA and RNA. Nevertheless, within most RNA molecules short, rod-like, double helical structures are formed by the single nucleotide chain folding back on itself.

DNA is the store of hereditary information in the cells, which is coded in the sequence of bases attached to the sugar phosphate chain. It is known that the DNA molecule acts as a template for synthesis of an RNA molecule – 'messenger RNA' – which then controls the synthesis of proteins at sites in the

Hydrogen-bonded base pairs

Adenine–thymine   Guanine–cytosine

cell called 'ribosomes'. Each group of three bases on the DNA molecule ultimately commands the performance of a particular operation in the protein synthesis, and each of the 64 possible combinations of three bases has a meaning – either the incorporation of a particular amino acid into the protein sequence or conclusion of chain extension – although some of the combinations duplicate the same command.

8.3 Nucleotide coenzymes

It is frequently found in cell reactions that, in addition to the enzyme and substrate, a third substance is required for reaction to occur. These compounds, known as **coenzymes**, are really reagents in the enzyme-catalysed reactions. It is often found that an enzyme has a highly specific requirement for one particular coenzyme and will not catalyse the cell reaction even in the presence of a substitute molecule of closely similar constitution (e.g. NAD^+ and $NADP^+$, p. 191). It seems reasonable to suppose that adsorption of the coenzyme onto the enzyme surface is a prelude to involvement in the enzyme-catalysed cell reaction, which would explain the enzyme–coenzyme selectivity.

Many of these coenzymes have nucleotide sections of the structure, which do not seem to be associated with the chemically active site of the coenzyme but may be involved in adsorption by the enzyme molecule. We will look at a selection of these coenzymes with an emphasis on the chemical aspects of their activity.

Many of these coenzymes are named after the nucleoside from which they are derived, and the names of the nucleosides are, in their turn, derived from the names of purine and pyrimidine bases (Table 8.2). The corresponding deoxyribonucleosides are called 'deoxyadenosine', etc.

190 Aromatic compounds, nucleic acids and nucleotide coenzymes

Table 8.2

Name of base	Name of the ribonucleoside
Adenine	Adenosine
Guanine	Guanosine
Cytosine	Cytidine
Uracil	Uridine
Thymine	Ribothymidine

8.3.1 Adenosine triphosphate (ATP)

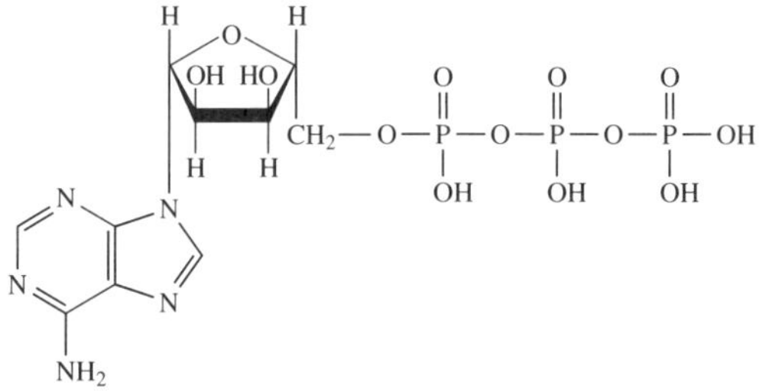

Adenine-D-ribose-phosphate-phosphate-phosphate*

ATP is universally distributed in living systems and represents a chemical store of energy in the cell. Reactions that, if left to proceed uncontrolled, would evolve large amounts of heat, are linked in living systems to other reactions that utilise the liberated energy in the synthesis of 'energy-rich' compounds (chemists normally refer to these as 'reactive' compounds). ATP is one of the most important of these compounds.

Biological reactions utilising ATP follow one of two paths. Either ATP acts as an acid anhydride and acylates the substrate (phosphorylation) or, more rarely, it can act as an alkylating agent. Numerous examples of phosphorylation are known, ATP being converted into the corresponding diphosphate – adenosine diphosphate (ADP) – by transfer of the terminal phosphate group. For example, D-glucose is phosphorylated to D-glucose-6-phosphate by ATP in the presence of the enzyme hexokinase:

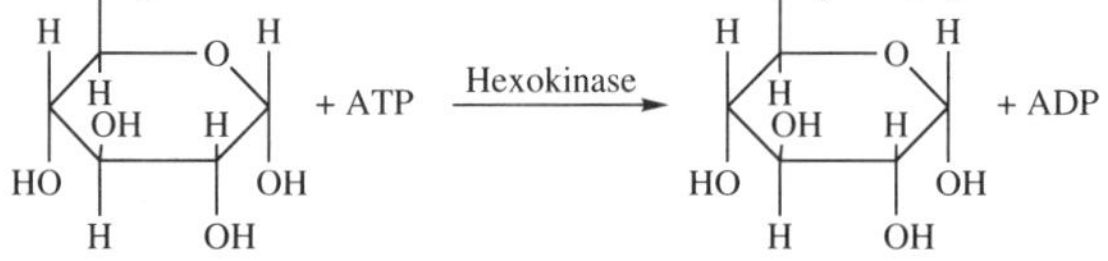

An example of ATP acting as an alkylating agent is in its reaction with the amino acid methionine:

*This and subsequent similar descriptions of nucleotides are not systematic names, but indicate how the complex molecules can be dissected into recognisable fragments.

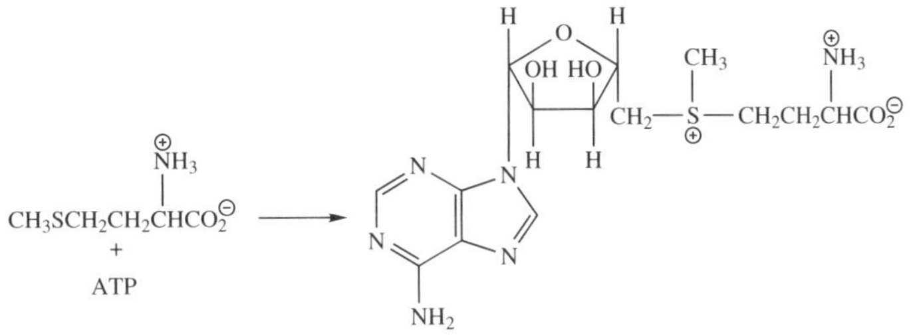

$$\text{CH}_3\text{SCH}_2\text{CH}_2\overset{\overset{+}{\text{NH}_3}}{\text{CH}}\text{CO}_2^{\ominus} + \text{ATP} \longrightarrow \text{[adenosyl-S}^+(\text{CH}_3)\text{-CH}_2\text{CH}_2\overset{\overset{+}{\text{NH}_3}}{\text{CH}}\text{CO}_2^{\ominus}\text{]}$$

The sulphonium cation obtained by this reaction is the biological methylating agent responsible for conversion of —OH and —NH— groups into —OCH$_3$ and —NCH$_3$— by transfer of the methyl group attached to the sulphur atom.

A closely related nucleotide coenzyme is adenosine-3′,5′-monophosphate ('cyclic AMP') which is formed from ATP by the enzyme adenylate cyclase:

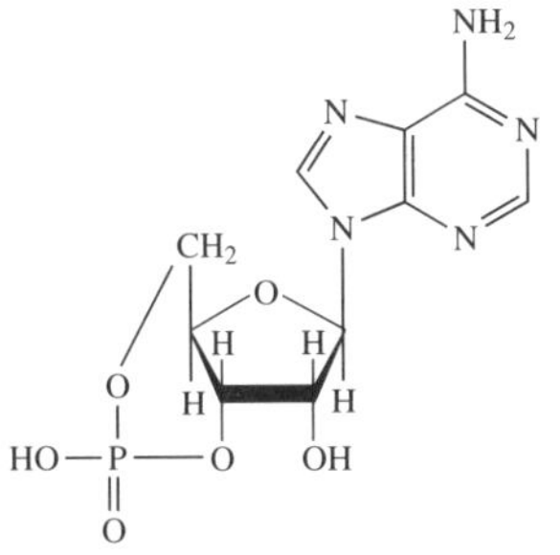

Cyclic AMP

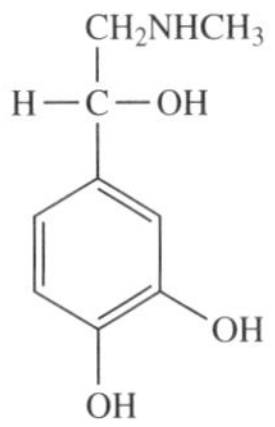

Adrenaline (epinephrine) (the natural compound has the *R* configuration)

This is an important compound involved in the regulation of cell reactions. A number of enzymes are known to exist in active and inactive forms, and the overall activity of the cell processes in which they are involved can be controlled by interconversion of the active and inactive forms. Cyclic AMP is required for the activation of a number of such enzymes, and is one of the compounds involved in a complicated sequence of steps linking adrenaline secretion with the stimulation of glycogen degradation and suppression of glycogen synthesis.

8.3.2 Nicotinamide-adenine dinucleotide (NAD$^+$) and nicotinamide-adenine dinucleotide phosphate (NADP$^+$)

These two coenzymes (formerly known as diphosphopyridine nucleotide, DPN, and triphosphopyridine nucleotide, TPN, respectively) are frequently involved in enzymic oxidations and reductions. The chemically active group in both cases is the nicotinamide moiety, which can undergo a reversible reduction in which the pyridinium ring is reduced to a dihydropyridine by

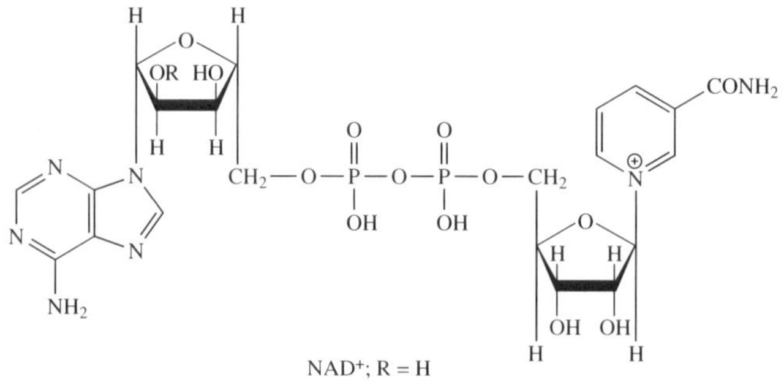

NAD⁺; R = H
NADP⁺; R = PO₃H₂

NAD = adenine-D-ribose-phosphate-phosphate-D-ribose-nicotinamide

reaction with a hydride ion or the chemically equivalent hydrogen ion and two electrons. These coenzymes and their reduced derivatives can therefore act as

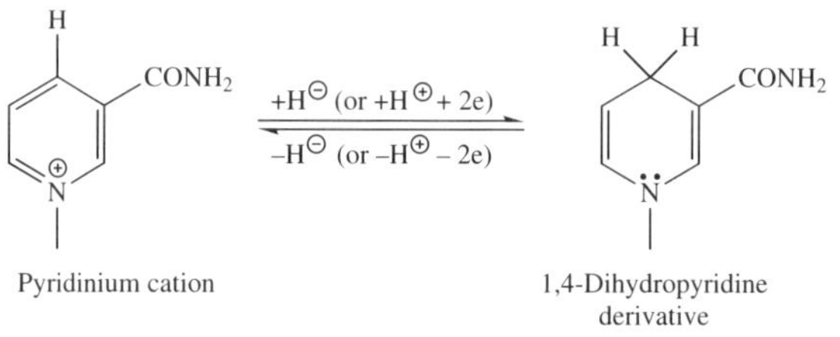

Pyridinium cation   1,4-Dihydropyridine derivative

acceptors or donors of H⁻ or electrons. The shorthand representation of these reactions is:

$$H^+ + 2e + NAD^+ \rightleftharpoons NADH$$

$$H^+ + 2e + NADP^+ \rightleftharpoons NADPH$$

Although the coenzymes differ only marginally in their structure, there is absolute specificity for one or the other in enzymic reactions. In general, degradative processes involve NAD⁺ and NADH, while synthetic processes utilise NADP⁺ and NADPH. The complete specificity for one or other coenzyme enables these processes to be controlled independently.

8.3.3 Flavine-adenine dinucleotide (FAD)

FAD is composed of a nucleotide unit joined to riboflavine (p. 178). This is another reduction–oxidation coenzyme existing in oxidised and reduced forms like NAD and NADP. If the alloxazine system (p. 177) is represented by the most 'aromatic' canonical structure, the reduction can be reasonably represented by a process similar to that known to occur with NAD.

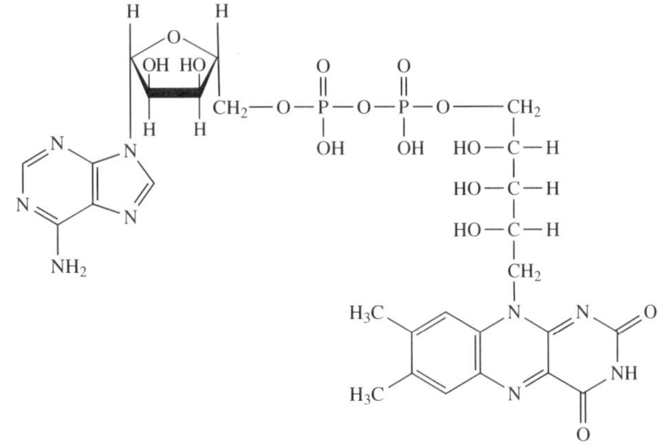

Adenine-D-ribose-phosphate-phosphate-riboflavine

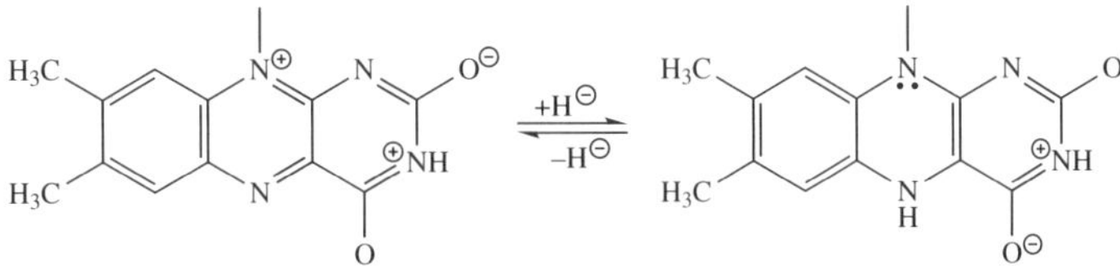

8.3.4 Coenzyme A (CoASH)

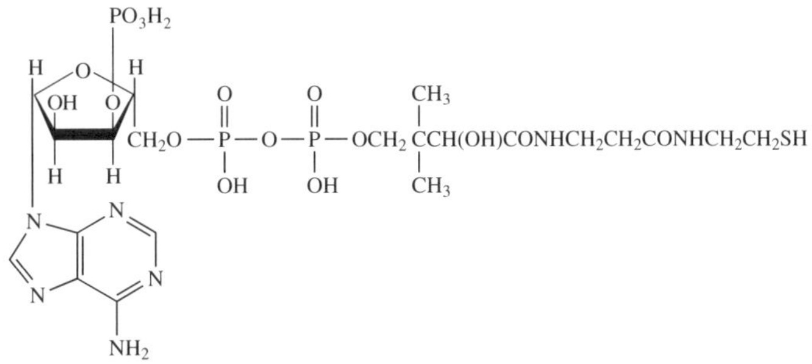

Adenine-D-ribose-(phosphate)-phosphate-phosphate-pantothenic acid-mercaptoethylamine

Despite the complexity of this coenzyme, which is widely involved in the metabolism of carboxylic acids, its chemical behaviour in enzymic reactions is that of a simple thiol.

8.4 Summary

1. Aromatic compounds are cyclic, planar and have $(4n + 2)$ delocalised π electrons. Anti-aromatic compounds are cyclic, planar and have $(4n)$ delocalised π electrons. Aromatic compounds are particularly stable and their characteristic reaction is electrophilic aromatic substitution (EAS), which preserves their aromaticity.

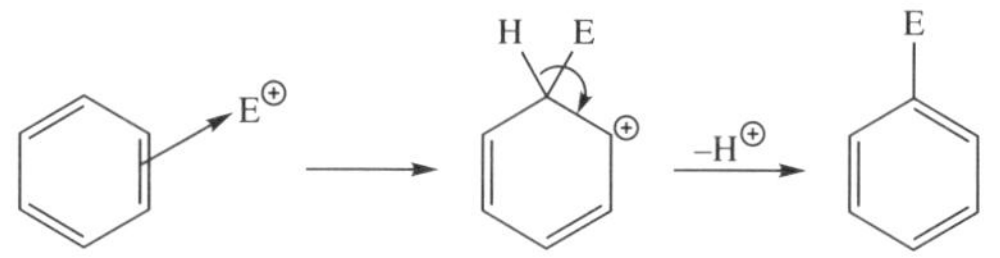

Anti-aromatic compounds are highly reactive and are unstable compared with their non-planar conformations.

2. Substituted benzene rings can be subdivided into three groups according to the effect of the substituent upon electrophilic aromatic substitution:

(I) Activating; *ortho/para* directing (e.g. CH_3, etc., $-NH_2$, $-OH$): these activate the benzene ring toward EAS, by making it more nucleophilic, and direct the incoming electrophile to the *ortho-* and *para-* positions by stabilising the carbocation intermediates for reactions at these positions, relative to that for *meta*.

(II) Deactivating; *meta* directing (e.g. CHO, CO_2R, NO_2, $\overset{+}{N}R_3$, $C\equiv N$, CX_3): these electron-withdrawing substituents reduce the electron density of the ring (and so deactivate it toward EAS) and destabilise the carbocation intermediates for *ortho-* and *para-* reaction, relative to *meta*.

(III) Deactivating; *ortho/para* directing (F, Cl, Br, I): although electron-withdrawing, and so deactivating, these substituents, which have lone-pairs, can stabilise the carbocation intermediates for *ortho-* and *para-* reaction by a mesomeric (resonance) interaction.

3. The nucleic acids RNA and DNA consist of chains of alternate sugar and phosphate residues, with heterocyclic bases attached to the sugar units:

$$—Phosphate—\underset{\underset{Base}{|}}{Sugar}—Phosphate—\underset{\underset{Base}{|}}{Sugar}—$$

RNA and DNA are distinguishable by the sugar unit they contain: D-ribose and 2-deoxy-D-ribose, respectively. In addition, the groups of bases also differ: both RNA and DNA contain adenine, guanine and cytosine, but RNA contains uracil and DNA contains thymine.

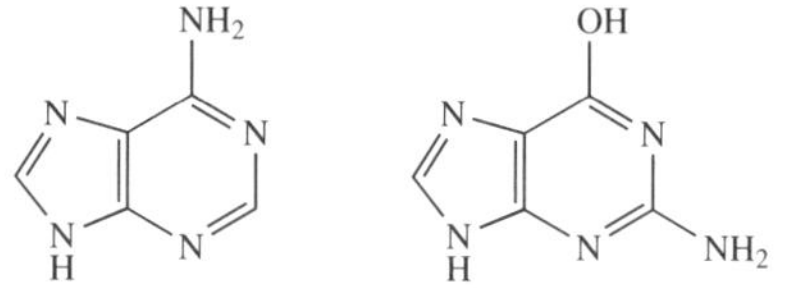

Adenine (A) Guanine (G)

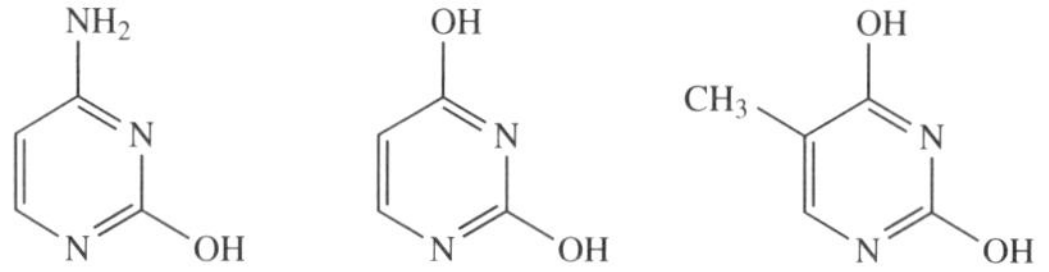

Cytosine (C) Uracil (U) Thymine (T)

DNA consists of a double helix with hydrogen bonding between adenine and thymine, or cytosine and guanine, in opposite strands.

Problems

8.1 Explain why pyrrole (**1**) is much less basic than pyridine (**2**) and why both compounds are classified as aromatic.

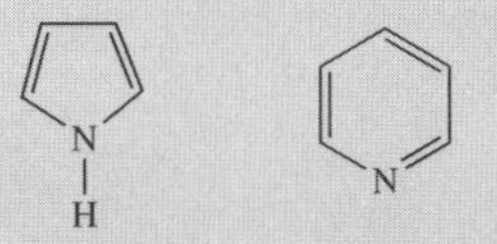

(1)   (2)

8.2 Which is the most basic nitrogen and which is the least basic nitrogen in histamine (**3**)? Give reasons for your choices.

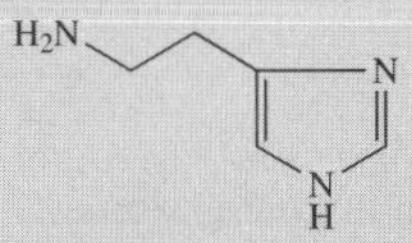

(3)

8.3 Show how benzene may be converted to ethylbenzene using the Friedel–Crafts process by:

(a) a single step route, and

(b) a two-step route.

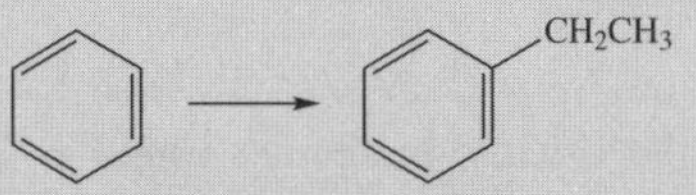

Discuss the mechanism of both processes and indicate the advantages that the two-step process may have in introducing longer chain alkyl groups, e.g. *n*-propyl, onto the benzene ring.

8.4 Show how benzene may be converted into the following compounds (give the reagents and any specific conditions required): (a) aniline, (b) chlorobenzene, (c) phenol, (d) toluene, (e) acetophenone.

8.5 On each of the following molecules indicate the most likely positions for substitution by an electrophile and indicate whether the molecules will be more or less reactive than benzene.

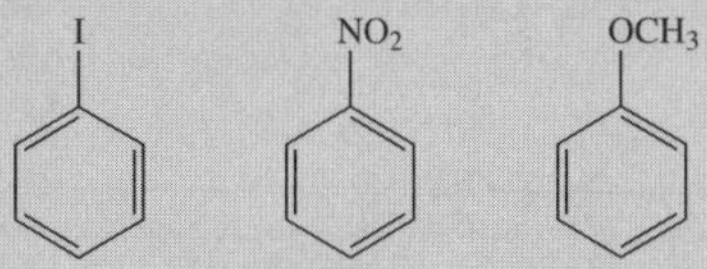

9 Lipids

Topics

9.1 Fatty acids
9.2 Plant and animal waxes
9.3 Depot fats
9.4 Phospholipids
9.5 Lipids and the structure of biological membranes
9.6 Summary

The smaller organic molecules found in living tissues can be divided into two broad groups. On the one hand there are the water-soluble materials such as amino acids and sugars which are insoluble in non-hydroxylic solvents such as chloroform or ether. The other group, comprising the fat-soluble materials which are soluble in chloroform, ether and other organic solvents, but are usually insoluble in water, are known by the generic name of **lipids**. Clearly, so crude a distinction as solubility in a particular group of solvents implies no specific structural feature in common, but within this broad range of substances are to be found many series of compounds with common functional groups and general constitutional similarities. The low solubility in water

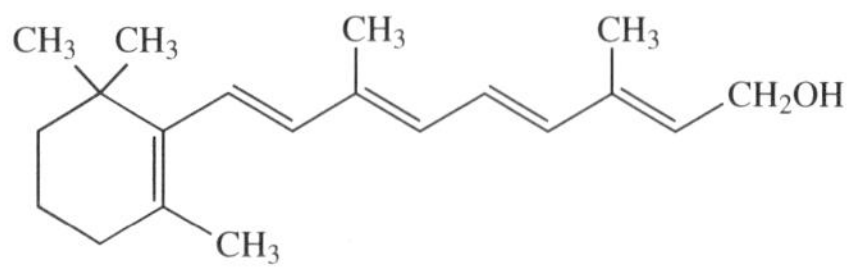

Vitamin A_1

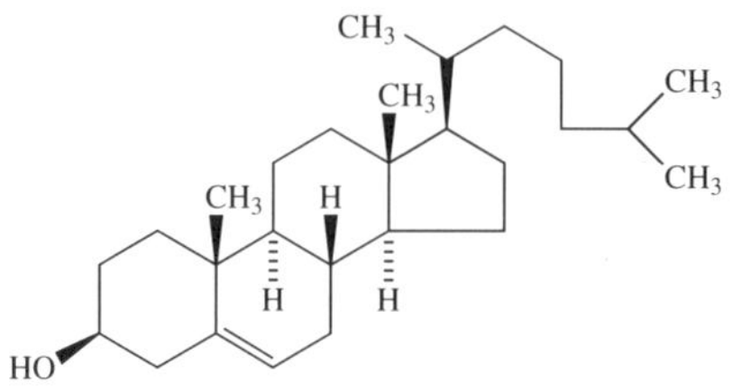

Cholesterol

implies that, in lipids, highly polar or hydrogen-bonding groups are either absent or constitute only a very small part of the whole molecule whereas non-polar (i.e. hydrocarbon) groups will predominate. Among the compounds included in the lipids are many of immense biological importance such as Vitamins A and D and steroids, e.g. cholesterol, which occur only in minute traces and collectively do not account for more than a tiny fraction of the total lipid content of any living system. Those constituents of the lipids that are most abundant are associated with very few general functions. One group of lipids function as the protective coatings on the cell walls of bacteria, the leaves of higher plants, the cuticle of insects and the skin of vertebrates.

The depot fats are another group, which form the store of metabolic fuel in living systems, and the third major group is that of the phospholipids, which are important components of biological membranes.

9.1 Fatty acids

Although the fatty acids (long-chain aliphatic carboxylic acids) occur in the free state only in trace amounts, they are one of the groups of simple molecules from which many lipids are constructed. The acyl moieties most

Table 9.1 Common fatty acids involved in lipid formation

No. of C atoms	Constitution	Systematic name	Trivial name
Saturated acids			
10	$CH_3(CH_2)_8CO_2H$	Decanoic acid	Capric
12	$CH_3(CH_2)_{10}CO_2H$	Dodecanoic acid	Lauric
14	$CH_3(CH_2)_{12}CO_2H$	Tetradecanoic acid	Myristic
16	$CH_3(CH_2)_{14}CO_2H$	Hexadecanoic acid	Palmitic
18	$CH_3(CH_2)_{16}CO_2H$	Octadecanoic acid	Stearic
20	$CH_3(CH_2)_{18}CO_2H$	Eicosanoic acid	Arachidic
22	$CH_3(CH_2)_{20}CO_2H$	Docosanoic acid	Behenic
24	$CH_3(CH_2)_{22}CO_2H$	Tetracosanoic acid	Lignoceric
26	$CH_3(CH_2)_{24}CO_2H$	Hexacosanoic acid	Cerotic
Unsaturated acids			
16	$CH_3(CH_2)_5CH=CH(CH_2)_7CO_2H$ Z-Hexadec-9-anoic acid		Palmitoleic
18	$CH_3(CH_2)_7CH=CH(CH_2)_7CO_2H$ Z-Octadec-9-enoic acid		Oleic
18	$CH_3(CH_2)_5CH=CH(CH_2)_9CO_2H$ E-Octadec-11-enoic acid		Vaccenic
18	$CH_3(CH_2)_4(CH=CHCH_2)_2(CH_2)_6CO_2H$ Octadeca-9(Z),12(Z)-dienoic acid		Linoleic
18	$CH_3CH_2(CH=CHCH_2)_3(CH_2)_6CO_2H$ Octadeca-9(Z),12(Z),15(Z)-trienoic acid		Linolenic
18	$CH_3(CH_2)_3(CH=CH)_3(CH_2)_7CO_2H$ Octadeca-9(Z),11(E),13(E)-trienoic acid		α-Eleostearic
20	$CH_3(CH_2)_4(CH=CHCH_2)_4CH_2CH_2CO_2H$ Eicosa-5(Z),8(Z),11(Z),14(Z)-tetraenoic acid		Arachidonic
24	$CH_3(CH_2)_7CH=CH(CH_2)_{13}CO_2H$ Z-Tetracosa-15-enoic acid		Nervonic

frequently encountered in the major lipid groups are those derived from straight-chain aliphatic acids with even numbers of carbon atoms, usually in the range C_{14}–C_{22} with C_{16} and C_{18} being the most abundant. Derivatives of fully saturated and mono- and polyunsaturated acids are found, but derivatives of carboxylic acids with C≡C groups are rare as are those of acids with branched chains or more elaborate structures. Amongst the unsaturated acids concerned, those with *cis* (*Z*)-stereochemistry about the double bond(s) are more common than the *trans* (*E*)-stereoisomers, and non-conjugated, polyunsaturated acids are more abundant than their conjugated isomers. Polyunsaturated acyl groups containing sequences of CH=CH—CH_2 groups are fairly common. Some of the commoner fatty acids involved in lipid formation are listed in Table 9.1.

9.2 Plant and animal waxes

These compounds, which form the protective, waterproof covering on many plants and animals, are usually the esters of long-chain fatty acids and alcohols with a large hydrocarbon group. Beeswax contains the palmitate esters of long-chain alcohols, and leaf waxes consist of esters of fatty acids and alcohols having up to 34 carbon atoms. Wool-wax, lanolin, contains the esters of lanosterol, a steroid.

$CH_3(CH_2)_{14}CO_2(CH_2)_{25}CH_3$ A component of beeswax

A component of wool-wax

In all these compounds, save for the two oxygen atoms of the ester function, the whole of the molecule consists of hydrocarbon groups. It is not very surprising that their physical properties resemble those of the larger hydrocarbons, e.g. paraffin waxes.

9.3 Depot fats

The depot fats, which form one of the metabolic fuel reserves of living systems, are predominantly triacyl derivatives of glycerol (p. 58). In general the triacylglycerols from animal sources differ from those of many plant oils in the higher proportion of saturated acyl groups in the animal fat. There is a clear correlation between the extent of unsaturation and the melting point in triacylglycerols, the highly unsaturated seed-oils having very low melting points, whereas animal fats are usually solid at ambient temperature.

The commercial catalytic hydrogenation of plant oils to form margarine results in the production of a commodity with physical properties resembling those of a typical animal fat. This difference in physical properties can be related to the different shapes of saturated and unsaturated fatty acid molecules, which is most clearly shown by considering the shapes of the molecules when the carbon chain is in the fully extended conformation.

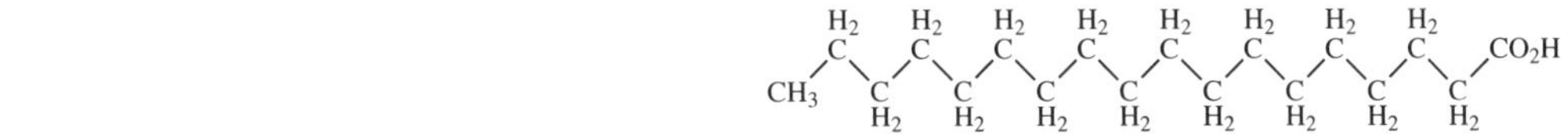

$CH_3(CH_2)_{16}CO_2H$ Stearic acid

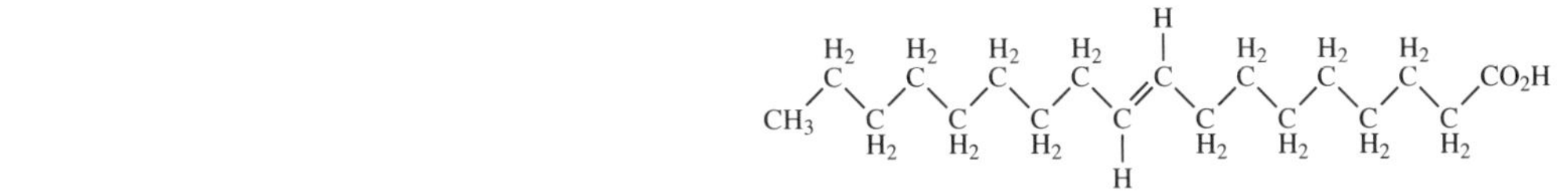

$CH_3(CH_2)_7CH=CH(CH_2)_7CO_2H$ Elaidic acid

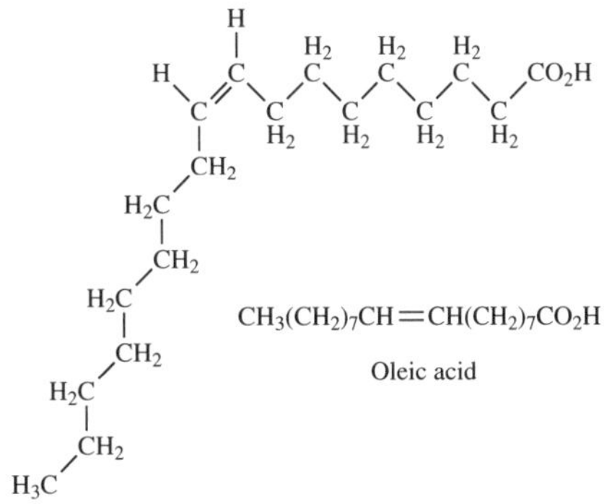

$CH_3(CH_2)_7CH=CH(CH_2)_7CO_2H$ Oleic acid

Whereas the *E* configuration about the double bond in elaidic acid produces scarcely any significant change in the molecular shape compared with the saturated analogue (stearic acid), the *Z* configuration of the unsaturated group in oleic acid produces a very pronounced kink in the molecule. Multiple unsaturation, such as in linolenic and arachidonic acids, will enhance the irregularity in the shape of an acyl chain so that more extensive unsaturation accompanied by *Z* stereochemistry at the double bonds will lead to more irregularity in the molecular shape of derivatives such as the triacylglycerols. Irregularity in molecular shape means that the molecules are less conveniently packed into a three-dimensional crystal lattice, which means a lower binding energy for the lattice and hence a lower melting point. In triacylglycerols and other lipids containing more than one long-chain acyl group, a wide variation in molecular shape and associated properties can be achieved by introduction of *Z*-unsaturation into the acyl hydrocarbon chain. It is not very surprising that fish-oils are more highly unsaturated than the body fats of warm-blooded animals since in the former case the requisite physical properties of the depot fats must be achieved at much lower temperatures. The incidence of coronary and arterial disease in humans also appears to be linked to the level of unsaturated fat in the diet, possibly for similar reasons.

9.4 Phospholipids

The phospholipids are a large group of fairly complex molecules occurring extensively in biological membranes and having a phosphate ester group as the common constitutional feature. The majority of phospholipids contain glycerol as one of the structural units and are derived from glycerol-1-phosphate (α-glycerophosphoric acid) with the R configuration at the chiral centre:

R-Glycerol-1-phosphate
(also called D-glycerol-1-phosphate
and L-glycerol-3-phosphate)

A phosphatidic acid

This is then acylated on the other two hydroxyl groups of the glycerol residue by long-chain fatty acids (it is often found that the 2-hydroxyl group of the glycerol residue is acylated by an unsaturated acid) to form 'phosphatidic acids' which are converted into the naturally occurring 'phosphoglycerides' (also called 'glycerophosphatides') by esterification of the phosphoric acid residue by another molecule containing a hydroxyl group. Several types of phosphoglycerides are known, distinguished by the nature of this final section of the molecule, which is always a highly polar or hydrogen-bonding group. Table 9.2 summarises the constitutional features of the more important types of phosphoglyceride.

A related group of phospholipids are the plasmalogens, differing from the phosphatidic acid derivatives by having an unsaturated ether group at position 3 of the glycerol residue in place of the normal acyl group:

$$CH_3(CH_2)_{15}CH=CH-OCH_2CHCH_2O-\overset{O}{\underset{OH}{\overset{\|}{P}}}-OCH_2CH_2\overset{\oplus}{N}(CH_3)_3$$
$$CH_3(CH_2)_7CH=CH(CH_2)_7CO-O$$

A plasmalogen

Glycolipids are constituents of plant membranes and although not phosphate esters are conveniently mentioned here. In glycolipids a 1,2-diacylglycerol is linked through the 3-hydroxyl group to a sugar, often D-galactose, which provides a highly hydrogen-bonding terminal group:

A glycolipid

Table 9.2
Types of phosphoglycerides

$$\begin{array}{c} \text{CH}_2-\text{CH}-\text{CH}_2-\text{O}-\overset{\overset{\displaystyle O}{\|}}{\underset{\underset{\displaystyle \text{OH}}{|}}{\text{P}}}-\text{X} \\ |\quad\quad | \\ \text{O}\quad\text{O} \\ |\quad\quad | \\ \text{CO}\quad\text{CO} \\ |\quad\quad | \\ \text{R}^1\quad\text{R}^2 \end{array}$$

$\quad\quad\quad\quad\quad\quad\quad\quad\quad\quad\quad\quad\quad\quad\quad$ —X

- Phosphatidyl ethanolamine (Cephalin): $\quad$ —OCH$_2$CH$_2$NH$_2$
- Phosphatidyl choline (Lecithin): $\quad$ —OCH$_2$CH$_2\overset{\oplus}{\text{N}}$(CH$_3$)$_3$
- Phosphatidyl glycerol: $\quad$ —OCH$_2$CHOH CH$_2$OH
- Phosphatidyl 3-*O*-aminoacylglycerol: $\quad$ —OCH$_2$CHOH CH$_2$O—CO—CH—R
$\quad\quad\quad\quad\quad\quad\quad\quad\quad\quad\quad\quad\quad\quad\quad\quad\quad\quad\quad\quad\quad\quad\quad$ |
$\quad\quad\quad\quad\quad\quad\quad\quad\quad\quad\quad\quad\quad\quad\quad\quad\quad\quad\quad\quad\quad\quad\quad$ $\overset{\oplus}{\text{NH}_3}$

- Phosphatidyl serine: $\quad$ —OCH$_2$ CH CO$_2^\ominus$
$\quad\quad\quad\quad\quad\quad\quad\quad\quad\quad\quad\quad\quad\quad$ |
$\quad\quad\quad\quad\quad\quad\quad\quad\quad\quad\quad\quad\quad\quad$ $\overset{\oplus}{\text{NH}_3}$

- Phosphatidyl inositol:

$$\text{Inositol} = \left[\text{inositol ring structure with OH groups}\right]$$

- Cardiolipin (1,3-bisphosphatidyl glycerol):

$$\begin{array}{c} \text{CH}_2-\text{CH}-\text{CH}_2-\text{O}-\overset{\overset{\displaystyle O}{\|}}{\underset{\underset{\displaystyle \text{OH}}{|}}{\text{P}}}-\text{O}-\text{CH}_2\text{CHCH}_2-\text{O}-\overset{\overset{\displaystyle O}{\|}}{\underset{\underset{\displaystyle \text{OH}}{|}}{\text{P}}}-\text{O}-\text{CH}_2-\text{CH}-\text{CH}_2 \\ |\quad\quad |\quad\quad\quad\quad\quad\quad\quad\quad\quad |\quad\quad\quad\quad\quad\quad\quad\quad\quad\quad\quad\quad\quad\quad |\quad\quad | \\ \text{O}\quad\text{O}\quad\quad\quad\quad\quad\quad\quad\quad\quad \text{OH}\quad\quad\quad\quad\quad\quad\quad\quad\quad\quad\quad\quad \text{O}\quad\text{O} \\ |\quad\quad |\quad\quad\quad\quad\quad\quad\quad\quad\quad\quad\quad\quad\quad\quad\quad\quad\quad\quad\quad\quad\quad\quad\quad\quad\quad\quad\quad\quad |\quad\quad | \\ \text{CO}\quad\text{CO}\quad\quad\quad\quad\quad\quad\quad\quad\quad\quad\quad\quad\quad\quad\quad\quad\quad\quad\quad\quad\quad\quad\quad\quad\quad \text{CO}\quad\text{CO} \\ |\quad\quad |\quad\quad\quad\quad\quad\quad\quad\quad\quad\quad\quad\quad\quad\quad\quad\quad\quad\quad\quad\quad\quad\quad\quad\quad\quad\quad\quad\quad |\quad\quad | \\ \text{R}^1\quad\text{R}^2\quad\quad\quad\quad\quad\quad\quad\quad\quad\quad\quad\quad\quad\quad\quad\quad\quad\quad\quad\quad\quad\quad\quad\quad\quad\quad \text{R}^3\quad\text{R}^4 \end{array}$$

Sphingolipids are derived from sphingosine or its dihydro-derivative sphinganine. Sphingosine is a long-chain molecule whose terminal three carbon atoms have a functionality resembling glycerol while the rest of the molecule has a long hydrocarbon chain similar to that of a fatty acid:

$$\text{CH}_3(\text{CH}_2)_{12}\text{CH}=\text{CH}-\underset{\underset{\displaystyle \text{OH}}{|}}{\text{CH}}-\underset{\underset{\displaystyle \text{NH}_2}{|}}{\text{CH}}-\text{CH}_2\text{OH}$$

Sphingosine
2(*S*)-Amino-1,3(*R*)-dihydroxyoctadec-4(*E*)-ene

In sphingolipids the 3-hydroxyl function of sphingosine or dihydro-sphingosine is usually unchanged and the amino group is acylated by a long-chain fatty acid. This 'ceramide' is then linked to a variety of groups through the terminal hydroxyl group:

$$\begin{array}{c} \text{OH} \\ | \\ \text{CH}_3(\text{CH}_2)_{12}\text{CH}=\text{CH}-\text{CH} \\ | \\ \text{CH}_3(\text{CH}_2)_{22}\text{CO}-\text{NH}-\text{CH}-\text{CH}_2\text{OH} \end{array} \qquad \text{A ceramide}$$

Sphingomyelins have a choline phosphate unit joined on to this terminal hydroxyl group (cf. lecithin), cerebrosides have a D-galactose residue attached here (cf. glycolipids), and gangliosides have complex oligosaccharide residues joined onto the 1 position of the ceramide:

$$\begin{array}{c} \text{OH} \\ | \\ \text{CH}_3(\text{CH}_2)_{12}\text{CH}=\text{CH}-\text{CH} \\ | \\ \text{R CO}-\text{NH}-\text{CH} \qquad \text{O} \\ | \qquad\qquad\quad \| \\ \text{CH}_2-\text{O}-\text{P}-\text{OCH}_2\text{CH}_2\overset{\oplus}{\text{N}}(\text{CH}_3)_3 \\ | \\ \text{OH} \end{array}$$

A sphingomyelin

9.5 Lipids and the structure of biological membranes

As previously stated, phospholipids, glycolipids and sphingolipids occur extensively in the membranes of living systems and are almost completely absent from depot fats. While the precise function of phospholipids, etc., in membranes is still unclear, it is well understood why these types of compound are found in association with that type of cellular structure. All of the types of lipid described above in Section 9.4 have molecular constitutions in which a large non-polar, hydrocarbon section of the molecule has a relatively small, polar or hydrogen-bonding portion. The general pattern is like a tuning fork with long, non-polar prongs and a short, highly polar leg. All the phospholipids have an acidic hydroxyl group on the phosphorus atom which will be dissociated at pH 7 to give a charged centre, and additionally cephalins, lecithins, phosphatidyl serine and sphingomyelins will be zwitterionic at ambient pH.

To understand why these compounds are involved in the formation of membranes it is necessary to look at some of the factors relating to the phenomenon of solubility. The extent to which a solute will dissolve in a solvent is determined by the relative strengths of solute–solute interaction in the solid state and solvent–solvent and solute–solvent interactions in the liquid phase. Polar solutes usually have strong forces binding the crystal lattice (e.g. electrostatic interaction in ionic or zwitterionic solids, or multiple hydrogen bonding as in the sugars), and they are unlikely to be easily dispersed in non-polar solvents where solute–solvent interactions will be weak and there will be little energy provided from this source to compensate for the energy required to remove molecules from the crystal lattice. Conversely, highly polar solvents are unlikely to dissolve non-polar solutes since insertion of non-polar solute molecules between polar solvent molecules would disrupt the relatively strong interactions between solvent molecules without any significant compensating solute–solvent interaction. So for those substances which dissolve as dispersed, isolated molecules there is a well-recognised qualitative relationship between the solubility and the relative polarities of solute and solvent.

A more complex situation exists where one considers the case of a molecule which contains both highly polar and non-polar regions. In a polar solvent, such as water, the solvent will interact with (solvate) the polar end of the solute, tending to cause this part of the molecule to dissolve, while the low solute–solvent interaction will tend to exclude the non-polar part of the molecule from the solvent medium. Such solutes, under these conditions, tend to form clusters known as **micelles** in which the non-polar ends of the solute molecule crowd together and the polar ends form an outer layer interacting with the polar medium. The net effect is to form a non-polar globule with a polar surface (Figure 9.1).

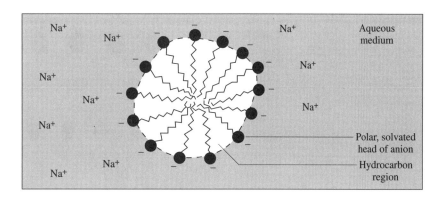

Figure 9.1

The alkali metal salts of long-chain fatty acids (soaps), e.g. Na^+ $^-O_2C(CH_2)_{16}CH_3$, are the most familiar example of this type of molecule, the polar end of the molecule being the carboxylate anion and the long hydrocarbon chain being the non-polar section. The detergent properties of these salts are due to the ability of the micelle to include grease and other non-polar material inside the hydrocarbon region, where it is, in effect, dissolved in a hydrocarbon solvent. The droplets are stabilised against coagulation by the surface electrical charge which repels any other approaching micelle with similar surface charge. In this way grease can be dispersed as an emulsion of droplets, stabilised in water by the soap. Long-chain quaternary ammonium salts (e.g. cetylpyridinium chloride, $CH_3(CH_2)_{15}\overset{+}{N}C_5H_5Cl^-$) form cationic detergents that act in precisely the same way.

The formation of micelles does not require an electrical charge on the solute. Whereas the triacylglycerols are insoluble in water, the mono- and diacylglycerols dissolve, forming micelles. In these cases the water-soluble (hydrophilic) section is the free hydroxyl group which can hydrogen bond to water and is responsible for the miscibility of glycerol and water. The water-insoluble (hydrophobic) sections in these glycerides are the long hydrocarbon chains of the fatty acid residues, as in the case of soaps.

Globular clusters are not the only type of structure formed by molecules with both hydrophilic and hydrophobic sections. Phospholipds and related compounds readily and spontaneously form monolayers on the surface of aqueous media (a), and bilayers in the aqueous media (b), the effect being like an extended micelle (Figure 9.2). The polar ends of the phospholipids are solvated by the water, and the hydrocarbon tails of the fatty acid residues, etc., form a non-polar, electrically insulating layer, impervious to the passage of charged species such as Na^+ and K^+.

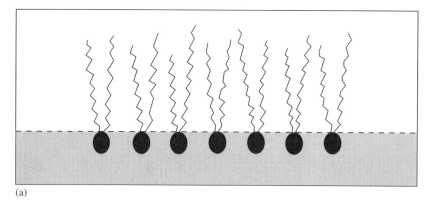

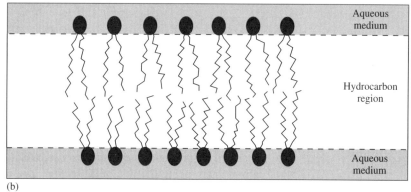

Figure 9.2

Biological membranes are not composed solely of phospholipid but contain on average 60 per cent protein and 40 per cent lipid, the lipid content including varying amounts of steroids, notably cholesterol. Despite the complexity of composition, the simple picture above is still valid since proteins also have non-polar and polar regions which can interact with the bilayer and aqueous medium respectively. There are apparently no covalent links between the protein and lipid constituents of the membranes, and individual molecules have some freedom of movement within the membrane. However, these lipoprotein complexes have considerable stability despite the absence of covalent bonding. How these membranes function in the chemistry of the cell is still far from clear and is well beyond the scope of this text.

9.6 Summary

1. Lipids are fat-soluble materials which are generally soluble in organic solvents but not in water. They play an important part in nature, for example, in the formation of biological membranes, as protective coatings, and as a store of metabolic fuel.

2. The plant and animal waxes are usually the esters of long-chain fatty acids and alcohols with a large hydrocarbon group:

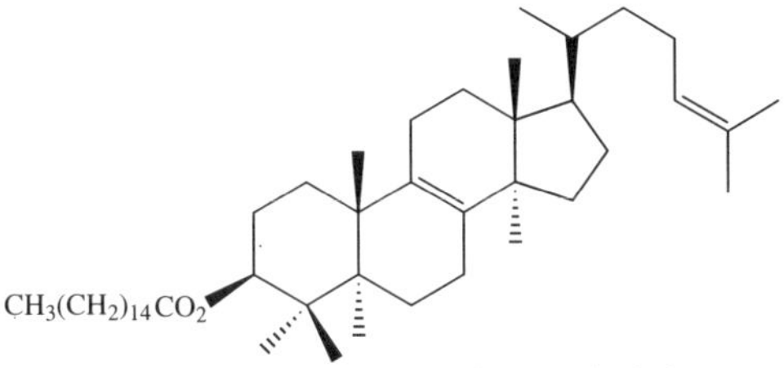

$CH_3(CH_2)_{14}CO_2$

Lanosterol palmitate

Depot fats are normally the triacyl derivatives of glycerol:

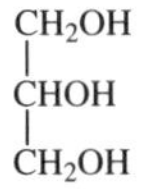

CH_2OH
$|$
$CHOH$
$|$
CH_2OH

Glycerol

The majority of phospholipids contain glycerol as one of the structural units and are derived from (R)-glycerol-1-phosphate:

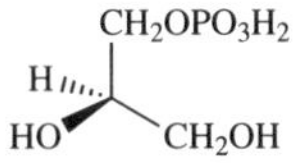

$CH_2OPO_3H_2$
H
HO CH_2OH

(R)-Glycerol-1-phosphate

3. Phospholipids have a non-polar hydrocarbon tail section as well as a relatively small, polar or hydrogen-bonding portion. In polar solvents, they form micelles or bilayers and it is this property that makes them ideal as components of membranes.

Index

acetaldelyde *see* ethanal
acetals, 89
acetic acid *see* ethanoic acid
acetone *see* propanone
acetophenone, 195
acetylcoenzyme A, 73
acetylene *see* ethyne
acetylsalicylic acid, 121
acyl halides, 111
addition
   electrophilic, 43
   to aldehydes, 86
   to alkenes, 43–8
adenine, 185
adenosine phosphates, ADP, ATP, cyclic AMP, 190–1
ajoene, 72
alanine, 161
alcohols
   biological preparation, 62
   dehydration, 71
   oxidation, 65, 104
   reaction with H_2SO_4, 66
   salt formation, 63
aldaric acids, 145
aldonic acids, 145
aldehydes
   addition of ammonia, 87
   addition of bisulphite, 90
   condensation reactions, 88
   reaction with nucleophiles, 85
aldolase, 94
aldol reaction, 93

aldoses, 143
alizarin, 96
alkenes
   addition to, 43–8
   formation, 50
   hydration, 60
   hydrobromination, 42
   oxidative cleavage, 83
alkoxides, 63
alloxazine, 177
alkyl halides
   formation, 65
   hydrolysis, 60
alkylbenzenes
   halogenation, 184
   oxidation, 104, 184
alkynes
   hydration, 84
amides
   dehydration, 118
   hydrolysis, 118
   oxidation, 129
   preparation, 110
   reaction as nucleophiles, 127
   reaction with HNO_2, 128
   reduction, 123
   salt formation, 117
amine oxides, 28
amines, 28
   reaction with HNO_2, 62, 68
amino acids, 124, 161
   essential, 160–1
   methylene-imines, 163

terminal, 164
aminopeptidase, 162
ammonium salts, 122, 130
amylase, 153
amylose, 152
aniline, 135, 195
anisole, 57
[18]-annulene, 175
anomers, 139
anthracene, 175
anti, 46
arginine, 161
arylhydrazones, 88
aromaticity, 172–4
atomic structure, 1
aufbau principle, 4
axial, 32
azo dyes, 133

base-pairs, 189
beeswax, 198
Benedict's reagent, 84
benzaldoxime, 35
benzene, 172
   resonance energy, 173
4,5-benzpyrene, 176
bisphenol A, 99
bond
   angles, 9–10
   covalent, 6
   semipolar, 6
   π, 7
   σ, 6
butadiene, 12
butane, 56
butan-1-ol, 77
but-2-ene, 33–4, 47

caffeine, 178
Cahn-Ingold-Prelog rules, 26–7, 55
camphor-10-sulphonic acid, 21–2
canonical structure, 13, 93
carbinols, 59
carbocations, 43–4, 49
carboxylic acids
   properties, 104
   reduction, 111

   salt formation, 109
carboxyl group, 102
cardiolipin, 201
β-carotene, 12
catechol, 95
cellobiose, 149, 156
cellulose, 151
ceramide, 202
chirality, 120
chlorobenzene, 195
chloroethanoic acid, 16
chlorocyclohexane, 32
cholesterol, 196
chymotrypsin, 165, 171
chromic acid, 65
cis, 16, 33
Clemmensen reduction, 85
coenzyme A, 117, 193
coenzymes, 189–190
configurations, 33
conformations
   boat, 30
   chair, 30
   eclipsed, 29
   cyclic compounds, 30
   staggered, 29
conjugate acid/base, 108
conjugation, 11, 38
covalency, 5
cyanohydrins, 86, 143
cyclobutadiene, 175
cyclobutane, 30
cyclohexane, 30
cyclooctatetraene, 175
cyclopentadienide anion, 175
cyclopentane, 30
cyclopropane, 30
   dicarboxylic acids, 36
cysteine, 161
cytidine, 190
cytosine, 185

2-deoxyribose, 185
detergents, 203
dextrorotatory, 17
diastereoisomers, 19, 21
diazonium salts, 129, 131–3

reduction, 132
1,4-dibromocyclohexane, 35
1,2-dibromoethane, 30
1,2-dibromoethene, 33
1,2-dichloroethene, 34
dihaloalkanes
   hydrolysis, 82
dihydroxypropanone, 99, 141
1,3-dimethylcyclohexane, 55
diols
   geminal, 88
   vicinal, 62
DIPAMP, 162
disulphides, 72–3
DNA, 186–8

E (configuration), 34
elaidic acid, 34, 199
electron
   shells, 2
   structure, 5
electronegativity, 37
electrophile, 43
electrophoresis, 168
electrovalency, 5
elimination, 50
enantianers, 20
enantiomerism, 16
enediol, 147
enolate anion, 93, 147
enols, 91
enovid, 54
enzymes, 189
epimers, 140
equatorial, 32
esters
   formation, 64
   hydrolysis, 61, 113
   preparation, 110
   reaction with NH_3, 114
   reduction, 114
ethanal, 80
ethane, 29
ethanediol, 58, 63, 64
ethanoic acid, 103
ethanol, 59
ethene, 10

ethers
   hydrolysis, 60
ethylene oxide, 77
ethyne, 11

FAD, 192
fatty acids, 197 (table)
Fehling's solution, 84
Fischer, 138
Fischer projection, 24–6
fission
   heterolytic, 39
   homolytic, 39
folic acid, 178
formic acid *see* methanoic acid
Friedel-Crafts, 195
   acylation, 84, 179
   alkylation, 179–80
fructose, 142, 156
furan, 139, 176
furanose, 139

galactose, 158
gentiobiose, 150
geometrical isomers, 16
glucitol, 145
gluconic acid, 145
D-glucosaccharic acid, 145
D-glucose, 156
   α-D-(+), 139
   β-D-(+), 139
   enolate, 147
   open-chair form, 140
   oxidation, 145
   pyranose forms, 140
   reaction with acetic anhydride, 146
   reduction, 145
   thioacetal, 146
glucoronic acid, 154
glucosamine, *N*-acetyl, 154
glutamic acid, 161
glutathione, 165–6, 171
glyceraldehyde, 23–4, 99, 141, 157
glycerol, 58, 62, 64, 78
   -1-phosphate, 200
glycogen, 153
glycolipid, 200

glycosides, 146
Grignard, 61, 94, 104
guanine, 185

halogenation
   of alkanes, 40
   of aromatics, 179–80
helix, 186–8
hemiacetals, 89
hexokinase, 190
hexoses, 137
histamine, 195
histidine, 160
Hofmann degradation, 125
Hog acylase, 23
Hund's rule, 5
hydrazones, 88
hydrobromination, 42
hydrogen, 4, 6
   bonding, 188–9
hydroquinone, 13

imidazole, 176
imines, 88
indole, 177
inductive effect, 37
insulin, 167
iodoform reaction, 101
ions, 5
isoelectric point, 160, 167
isoleucine, 161
isomers, 13
   geometrical, 16
   optical, 16
isoquinoline, 177
isotopes, 2, 3

Jones' reagent, 81

ketals, 90
keto-enol tautomerism, 91
ketones
   addition reactions, 86
   condensation reactions, 88
   reaction with, ammonia, 87
      bisulphite, 90
      nucleophiles, 85

ketoses, 142
Killiani-Fischer synthesis, 143

lactose, 151, 156
laevorotatory, 17
lanolin, 198
leucine, 161
lipid bilayers, 204
lithium aluminium hydride, 85
lysine, 161

maltose, 149, 156
mannose, 144
Markownikoff's rule, 45
membranes, 204
meso isomers, 26
mesomeric
   effect, 38
   structures, 13
methane, 9
methanoic acid, 103
methanol, 59
methionine, 161
methyl glucosides, 146
mevalonic acid, 22
micelles, 203
molecular structure, 8
monosaccharides, 137
Monsanto process, 162
muramic acid, N-acetyl, 154
mutarotation, 139

NAD, NADP, 82, 191
naphthalene, 175
neutrons, 2
Newman projections, 29
nicotinamide, 177
ninhydrin, 163
nitration, 179–80
nitro group reduction, 185
nitrophenols, 77
nomenclature *see also* individual classes
      of molecules
   alkanes, 14
   alkenes, 15
   alkynes, 15
nonylphenol, 99

nucleophiles, 48–9
nucleoprotein, 185
nucleosides, 186
nucleotides, 186
nucleus, 1, 2

oleic acid, 34, 199
optical isomers, 16
orbitals, 3
   atomic, 3
   hybridisation, 8
   molecular, 5, 6
   p, 3
   π, 7
   sp, sp^2, sp^3, 9–11
   σ, 7
osazones, 144
oxidation
   alcohols, 65, 81–3
   alkenes, 83
   polyols, 67
   vicinal diols, 83
oximes, 88
oxytocin, 166, 171

papain, 165
paracetamol, 135
Pauli exclusion principle, 5
pentan-2-ol, 58
pentoses, 137
pepsin, 165, 171
peptide linkage, 164
periodic acid, 67
pH, 105
phenanthrene, 175
phenols, 67, 195
   preparation, 68
   reactions, 69
phenylalanine, 161
phenylglycine, 162
phosphatidic acid, 200
phosphoglycerides, 201 (table)
phosphoketoepimerase, 148
phosphopentose isomerase, 148
phthalic anhydride, 20
piperidine, 32
pKa, 106–7 (table)

plasmalogen, 200
polarised light, 17
polysaccharides, 137
proline, 160
propanediol, 78
propanone, 81
proton, 2
protein structure, 168
pteridine, 177
purine, 177
pyran, 139
pyrazine, 176
pyrene, 176
pyridazine, 176
pyridine, 176, 195
pyridoxal phosphate, 124, 129, 177
pyridoxamine phosphate, 130
pyrimidine, 176
pyrrole, 176, 195

quantum number, 2
quaternary ammonium salts, 122–3, 130–1
quinoline, 177

R (chirality), 26
racemic, 17
radicals, 39, 40–3
   allyl, 42
   benzyl, 42
   stability, 42
   structure, 42
reactions
   electrophiles, 43–8
   nucleophiles
   radicals, 40–3
resolution
   biological, 22
   mechanical, 18
   via diastereoisomers, 19
resonance, 13, 91
   energy, 173
   hybrid, 13, 91
riboflavin, 178
ribose, 141, 158, 185
Ruff degradation, 145
Ruhemann's purple, 163

S (chirality), 26
salsolin, 78
Sandmeyer, 132
Schiff's bases, 88, 125, 127
semicarbazones, 88
semipolar bond, 6
serine, 161
silk fibroin, 169
S_N1, S_N2, 48–9
sorbitol, 145
sorbase, 158
specific rotation, 18
sphingolipids, 201
sphingomyelin, 202
sphingosine, 201
starch, 152
   degradation, 153
stearic acid, 199
stereoisomers, 16
Strecker synthesis, 161
substitution
   electrophilic, 48, 57, 70, 131, 179
     orientation in, 181
      substituent effects, 182–4
   nucleophilic, 48–50
sucrose, 151, 156
sulphonic acids, 68, 179, 181
sulphonium cations, 28
sulphoxides, 28

tartaric acid, 26, 55
tautomerism, 91
tetrahydrofolic acid, 178
tetrahydropyran, 32
tetroses, 137
theobramine, 178

theophylline, 178
thiamine pyrophosphate, 177
thiazole, 176
thioesters, 116
thioethers, 72–4
thiophene, 176
thiols, 72–4
thymine, 185
thyroxine, 70
Tollen's reagent, 84
trans, 16, 33
transesterification, 113
trehalose, 156
trioses, 137
tropylium catian, 175
trypsin, 165
tryptophan, 160
tyrosine, 160

uracil, 185
uric acid, 178

vitamin A_1, 196
vitamin K_1, 96
valency, 5
   coordinate, 6

Walden inversion, 49
Wheland intermediate, 179
Williamson synthesis, 71
Wolff-Kishner reduction, 85

Z (configuration), 34
Z-enzyme, 153
zwitterions, 159